n Introduction to Digital Logic

Other titles in Electrical and Electronic Engineering

J. C. Cluley: Electronic Equipment Reliability

R. F. W. Coates: Modern Communication Systems

W. Gosling: A First Course in Applied Electronics

B. A. Gregory: An Introduction to Electrical Instrumentation

Paul A. Lynn: An Introduction to the Analysis and Processing of Signals

A. G. Martin and F. W. Stephenson: Linear Microelectronic Systems

J. E. Parton and S. J. T. Owen: Applied Electromagnetics

J. T. Wallmark and L. G. Carlstedt: Field-Effect Transistors in Integrated Circuits

An Introduction to Digital Logic

A. Potton
Principal Lecturer
School of Electronic and Electrical Engineering
City of Leicester Polytechnic

© A. Potton 1973

All rights reserved. No part of this publication may be reproduced or transmitted, in any form or by any means, without permission.

First edition 1973
Reprinted with corrections 1975

Published by
THE MACMILLAN PRESS LTD
London and Basingstoke
Associated Companies in New York Dublin
Melbourne Johannesburg and Madras

SBN 333 14679 4

Printed in Great Britain by
Tinling (1973) Ltd., Prescot, Merseyside.

This book is sold subject to the standard conditions of the Net Book Agreement.

The paperback edition of this book is sold subject to the condition that it shall not, by way of trade or otherwise, be lent, resold, hired out, or otherwise circulated without the publisher's prior consent, in any form of binding or cover other than that in which it is published and without a similar condition including this condition being imposed on the subsequent purchaser.

Contents

Preface ix

1 COMBINATIONAL LOGIC CIRCUITS 1

1.1 Combinational binary logic circuits 2
1.2 The NOT gate 3
1.3 Truth tables 5
1.4 Positive and negative logic 5
1.5 Noise immunity 6
1.6 NAND and NOR gates 7
1.7 The NOR operation 8
1.8 The NAND operation 9
1.9 The NAND–NOR relationship 10
1.10 AND and OR gates 11
1.11 The AND operation 11
1.12 The OR operation 11
1.13 The AND–OR relationship 12
1.14 Interconnection of gates 13
Problems 15

2 BOOLEAN ALGEBRA 17

2.1 George Boole 17
2.2 Boolean variables 18
2.3 Boolean inversion 18
2.4 The boolean AND operation 18
2.5 The boolean OR operation 18
2.6 The NAND operation 18
2.7 The NOR operation 18
2.8 Boolean expressions and truth tables 19
2.9 Simplification of boolean expressions 20
2.10 Summary of boolean relationships 21
Problems 22
Bibliography 23

3 DESIGN OF SIMPLE LOGIC SYSTEMS — 24

- 3.1 A simple design example — 24
- 3.2 An additional variable — 25
- 3.3 A second design example — 27
- 3.4 An equivalence system — 30
- Problems — 32

4 KARNAUGH MAPS — 34

- 4.1 Looping two cells — 36
- 4.2 The Karnaugh map technique — 37
- 4.3 Looping larger groups of cells — 38
- 4.4 Further cell looping techniques — 39
- 4.5 Looping cells containing 0's — 41
- 4.6 Karnaugh maps for NAND–NOR logic — 41
- 4.7 Design example — 43
- 4.8 A second example — 45
- 4.9 'Don't care' situations — 46
- 4.10 Minimisation with functions of more than four variables — 48
- Problems — 48
- Bibliography — 49

5 BISTABLE SYSTEMS — 50

- 5.1 The R–S flip–flop — 50
- 5.2 Discrete component circuits — 52
- 5.3 The clocked R–S flip–flop — 53
- 5.4 Counting elements — 54
- 5.5 The type D flip–flop — 54
- 5.6 The master–slave technique — 55
- 5.7 The J–K flip–flop — 55
- 5.8 Integrated circuit bistable elements — 57
- Problems — 57
- Bibliography — 59

6 COUNTERS AND REGISTERS — 60

- 6.1 Storage and shift registers — 60
- 6.2 Parallel and serial input and output of data — 62
- 6.3 Ripple-through counters — 63

6.4	Frequency division	63
6.5	Binary counters	64
6.6	Modification of the counting period	66
6.7	A scale of six binary counter	66
6.8	A scale of five counter	67
6.9	A scale of twelve counter	68
	Problems	70
	Bibliography	70

7 SYNCHRONOUS COUNTERS 71

7.1	A simple two-stage counter	71
7.2	A four-stage counter	72
7.3	Variable sequence counters	78
	Problems	81
	Bibliography	82

8 SIMPLE SEQUENTIAL LOGIC SYSTEMS 83

8.1	An electronic combination lock	83
8.2	Lift control systems	85
8.3	Timing sequences	88
8.4	Integrated circuit decoders	91
8.5	Decoding noise	92
	Problems	94

9 BINARY ARITHMETIC OPERATIONS 96

9.1	Binary adders	96
9.2	Multidigit parallel adders	99
9.3	Carry-bypass adders	101
9.4	Serial adders	102
9.5	Signed binary numbers	103
9.6	Generation of complements	104
9.7	Binary subtraction	106
9.8	Binary multiplication	107
9.9	Binary division	109
9.10	The rate multiplier	113
	Problems	116
	Bibliography	117

10 PRACTICAL CONSIDERATIONS 118

10.1 Integrated circuit families 118
10.2 Packaging of digital integrated circuits 118
10.3 Practical system development 119
10.4 Digital circuit parameters 120
10.5 Logic levels 120
10.6 Threshold levels 120
10.7 Noise margin 120
10.8 Fan out 121
10.9 Propagation delay time 121
10.10 Wired logic 122
10.11 Resistor-transistor logic (R.T.L.) 122
10.12 Diode-transistor logic (D.T.L.) 123
10.13 Transistor-transistor logic (T.T.L.) 124
10.14 Emitter coupled logic (E.C.L.) 125
10.15 Metal oxide semiconductor integrated circuits (M.O.S.) 126
10.16 Complementary MOS logic (C.M.O.S.) 128
Bibliography 131

APPENDIXES 132

Appendix A Systems of Numeration 132
Appendix B Logic Symbols 140

INDEX 143

Preface

During the past few years, we have seen an unprecedented expansion in the applications of electronic techniques. It is no accident that a large part of this expansion has involved systems using digital techniques. Undoubtedly, economic factors provide the major reasons for this. Digital systems which, even a few years ago, would occupy weeks of an engineer's time in the design phase can now be bought as low cost, 'off the shelf' integrated circuits. As a result of this the engineer can now contemplate the design of systems which would previously have been considered totally uneconomic and hopelessly complex. A further consequence of the expansion of digital electronic techniques has been that engineers in fields not traditionally associated with electronics have become aware of the potentialities and power of these techniques.

The aim of this book is to introduce the reader to electronic digital systems and assist him to develop the techniques necessary for the understanding and design of such systems. The book is essentially an introductory text, no previous knowledge of digital systems or indeed of general electronics being assumed. From time to time in the text, reference is made to the discrete component circuit form of some digital system elements. Particular examples are the NOT and NOR gates in chapter 1 and the R-S bistable circuit in chapter 5. The brief explanations which accompany these circuits are intended to assist readers who have some background knowledge of electronic circuits and devices. Such a background is not essential to the general understanding of the text however and readers who have no previous knowledge of the subject may accept the description of the characteristics of the circuit without being unduly worried about the actual details of the circuit action.

The book should prove suitable for use as a text for first and second year HND and BSc engineering students at Polytechnics, Universities and Technical Colleges. Much of the material will also be useful to students pursuing the more advanced stages of some technician level courses such as the HNC in electrical engineering.

The common types of logic gates are introduced in chapter 1 of the text as are the ideas of positive and negative logic. The action of gates and logic systems is described in this chapter mainly by the use of truth tables.

Chapter 2 provides a brief introduction to boolean algebra. In a text of this kind, it is neither essential nor desirable to include a detailed treatment of boolean algebra with full mathematical rigour. The object of chapter 2 is to establish the basic concept of two-state variables and accustom the reader to the boolean manner of describing logical relationships. A list of some useful boolean relationships is included at the end of this chapter for completeness. There is no suggestion that these relationships should be committed to memory before proceeding with the remainder of the text.

Chapter 3 is perhaps unusual in that it discusses design techniques before the reader has been introduced to logical map methods. The purpose of this is twofold;

firstly to provide the reader with an incentive to persevere with the study of the techniques which assist the design of logic systems, and secondly, to accustom the reader to the use of boolean terminology.

Logical maps are dealt with in chapter 4 and are used in the remainder of the text when simplification of a logical function is required.

Chapters 5 to 8 cover the more formal aspects of sequential logic systems starting with bistable elements in chapter 5, proceeding to registers and ripple-through counters in chapter 6 and synchronous counters in chapter 7. Chapter 8 deals with a selection of practical design problems which require the use of sequential logic elements for their solution.

In chapter 9 systems are described for performing arithmetic operations. It was felt that a chapter of this type merited inclusion if only because an introductory text on logic design is often used as a preliminary to a more detailed study of computer hardware.

Chapter 10 contains a review of important logic element parameters. Also included in this chapter are details of the important ranges of currently available integrated circuits. As has been said previously, the reader with no formal electronics background may choose to pass over the details of the circuit operation of particular logic families although the importance of the various parameters should be understood.

My grateful thanks are extended to Dorothy, without whose understanding, tolerance and superlative typing this book would never have been finished. In conclusion, the author would welcome any useful comments or criticism on the material presented in this modest text.

Alan Potton

1

Combinational Logic Circuits

In the study of engineering, a distinction is usually drawn between two types of systems, identified by the adjectives linear and digital respectively. In a book whose purpose is a study of systems which form a subdivision of one of these categories, it is appropriate to commence by considering the essential differences between the two types of systems. Of course, as is the case with many such attempts at rigid classification, the distinction between digital and linear systems is not as sharp as is implied, since many systems include both linear and digital elements.

In linear systems, the observable physical variables such as pressure, voltage, current, angular position, etc., will usually be able to change in a continuous manner, that is, not in abrupt steps. More important however is that we are prepared to recognise, at least theoretically, any change, however small. The system itself will of course place an upper limit on the magnitude of the changes.

In contrast, in digital systems the possible values of the physical quantities are divided into categories or classes. To take an example, a digital system which includes a rotating shaft might divide the possible positions of the shaft into ten classes, each class encompassing a $36°$ range of angular position. If the position of the shaft lies between 0 and $36°$ then it is in class 1, between $36°$ and $72°$ it is in class 2 and so on. Changes in the angular position of the shaft are not regarded as significant until they are of sufficient magnitude to take the variable, that is the shaft position, into a different class.

When describing digital systems, it is often convenient to imagine the physical variables changing from class to class in an abrupt manner although of course truly instantaneous changes can never take place in any physically realisable system. What happens in practice is that in most digital systems, the physical variables change fairly rapidly when moving between classes and remain relatively stable when in the class. The speed of the change may be very high indeed in electronic systems—as short as 1 nano second (10^{-9} seconds) with the fastest elements currently available. Irrespective of how the variables change, however, it is the class which the value of a variable occupies at any one time which is important.

The number of classes which we allocate to the physical quantities is often determined by the character of the devices or elements used in the system. Among the most successful of early types of digital system for example, were the mechanical and electromechanical calculating machines. These machines mostly identified ten classes, that is, they were decimal digital systems. It is perhaps unfortunate that most electronic devices do not lend themselves to systems in which the possible values of the physical quantities, voltage and current in this case, can easily be divided into ten categories. Various attempts have been made to produce digital electronic circuits which operate in a decimal mode but with little commerical success. With only very few exceptions, electronic digital systems have identified two classes only in the description of voltage or current, that is, they are binary systems. This is mainly because the transistor (which is the most important electronic device) can be operated in two stable and well-defined states known as cut-off and saturation. These two states are in fact normally avoided when transistors are incorporated in linear electronic circuits. One of the very few electronic systems which does not have two classes is the range of logic elements referred to as tri-state logic. These devices do not exactly divide the total voltage range into three classes, nevertheless, three states are distinctly definable. Although these devices are described in chapter 10, the reader is advised at this stage to concentrate on the electronic logic system elements which are of major importance—binary logic circuits.

When discussing digital logic systems, it is possible to describe a particular system as either 'sequential' or 'combinational'. If a system is combinational it is always possible at any time to predict the output of the system if the inputs to the system are known at that time. This is not the case with sequential systems however since at any time, the output depends not only on the input at that time but also on the value of input(s) at some previous time interval. Sequential logic systems usually contain some combinational logic elements and sequential logic elements can themselves be formed from combinational elements as described in chapter 5.

Two types of sequential logic systems may be encountered. The two types are known as asynchronous and synchronous systems. Asynchronous sequential logic systems are also sometimes referred to as ripple-through systems. In asynchronous systems, any changes in the outputs which are initiated by changes in the inputs occur immediately the inputs change (after the normal delays within the system elements of course). Synchronous systems, in addition to the normal inputs which control the changes occurring within the system and at the outputs, have an additional 'timing' control often called the 'clock' input. Changes can only take place within the system and therefore at system outputs when the clock input is activated, although the actual changes which do then occur are determined by the remaining system inputs. The distinction between synchronous and asynchronous systems is important and the design methods differ considerably for the two types of system as will be seen in chapters 6 and 7.

1.1 Combinational Binary Logic Circuits

Electronic binary logic circuits may be loosely described as circuits in which only two sorts of voltage level are recognised—'high' voltages and 'low' voltages. The way in which a particular value of voltage falls into one category or the other will depend on the particular type of logic circuitry under consideration. Because of the fact that the precise value of a voltage is seldom important and it is only whether it can be described as 'high' or 'low' which matters, tolerance on component values

COMBINATIONAL LOGIC CIRCUITS

and active device parameters can be very wide. This leads to low manufacturing cost, easy application of production line techniques and high reliability compared with other types of electronic circuitry.

1.2 The NOT gate

This is the simplest type of logic gate and provides a suitable introduction to the basic ideas of logic circuits. The symbol for this type of gate is shown in figure 1.1 and the relationship between the input and output voltages for a typical gate is shown in figure 1.2.

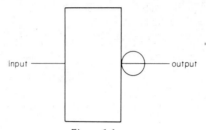

Figure 1.1

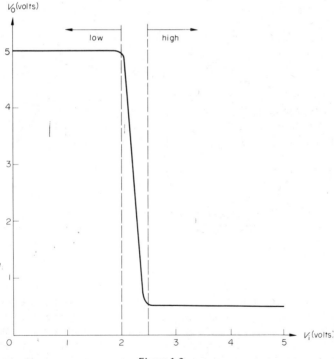

Figure 1.2

For a gate with the transfer characteristic shown in figure 1.2, all voltage levels below 2 V would be classified as 'low' and all voltage levels above 2.5 V would be classified as 'high'. Voltage levels between 2 V and 2.5 V are not normally regarded as identifiable since they do not fall into either the 'low' or the 'high' category. It is generally desirable in logic circuits that the range of such non-defined voltages should be as small as possible.

A circuit for a NOT gate which uses discrete components is shown in figure 1.3.

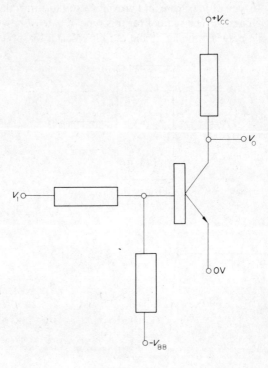

Figure 1.3

Figure 1.4 shows the transistor output characteristics with the load line superimposed in the normal manner. When the input voltage is zero, the transistor is held cut off with the operating point at X, the output voltage being approximately V_{cc}. As the input voltage is made more positive, a point will be reached when the base voltage of the transistor reaches the turn-on voltage and the transistor will start conducting. As the input voltage continues increasing, the operating point moves up the load line to Y. At this point, the transistor is saturated and the output voltage remains approximately constant at $V_{CE(SAT)}$ as the input voltage increases further. It will be seen that the input-output voltage relationship is of the same form as that shown in figure 1.2. Discrete component logic circuits are becoming relatively unimportant as the use of integrated circuit techniques becomes almost universal. Particularly for the engineer accustomed to the more traditional techniques, however, an examination of the behaviour of a discrete component version can sometimes help illuminate the action of a particular type of gate.

COMBINATIONAL LOGIC CIRCUITS

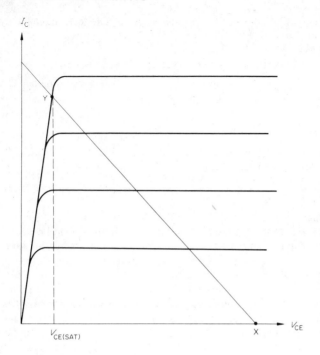

Figure 1.4

From the transfer characteristic shown in figure 1.2 we could describe the operation of a NOT gate by the following statement:

When the input voltage is 'low', the output voltage is 'high' and when the input voltage is 'high', the output voltage is 'low'.

1.3 Truth tables

Such a description of the circuit operation is necessarily verbose and as a result, various symbolic methods of describing logic circuit behaviour have been devised. One of the most common techniques is the *truth table* method. Table 1.1 shows a truth table for the NOT gate.

Table 1.1

input	output
L	H
H	L

1.4 Positive and negative logic

Rather than using the terms 'high' and 'low', most workers use the symbols **1** and **0** to describe the inputs and outputs of logic circuits. Two alternatives are possible giving rise to *positive logic* or *negative logic* systems.

(1) **1** represents a 'high' voltage
   0 represents a 'low' voltage

This designation is referred to as *positive logic*. The truth table for the NOT gate using this nomenclature is shown in table 1.2.

Table 1.2			Table 1.3	
input	output		input	output
0	1		1	0
1	0		0	1

(2) 1 represents a 'low' voltage
    0 represents a 'high' voltage

This designation is referred to as *negative logic*. The truth table for the NOT gate is shown in table 1.3.

It is apparent that the two truth tables are in fact identical in substance. One might think from this that the distinction between positive and negative logic is trivial but it can become significant when more complicated gating functions are investigated. Although the symbols 1 and 0 can be interpreted in two different ways giving use to positive and negative logic, it must also be remembered that the terms 'high' and 'low' to classify voltage levels are not completely unambiguous. It has become conventional to regard 'low' voltages as the least positive (or most negative) voltages and 'high' voltages as the most positive (or least negative) voltages. Taking as an example a circuit in which voltage levels could be classified as lying in the range 0 to -5 V or -6 V to -12 V, voltages in the 0 to -5 V range would be classified as 'high' and voltages in the -6 V to -12 V range as 'low'. On the basis of tables 1.2 and 1.3, the operation of the gate could be described by the statement

when the input is 1, the output is not 1

1.5 Noise immunity

Returning now to the gate transfer characteristic shown in figure 1.2, it will be seen that for input voltage levels above 2.5 V the output level is about 0.5 V which is in the 'low' voltage category. Suppose now that the output of this gate were used as the input to a second gate. The output of the second gate would be 'high' because the input was 'low'. If this second gate were in a situation in which electrical interference or noise produced voltage spikes on the input line, it is fairly obvious that the amplitude of such spikes would have to exceed 1.5 V (that is, sufficient to raise the voltage level from 0.5 V to above 2 V) if any effect were to be observed on the output. A similar situation would exist if the normal 'high' output level of 5 V of a gate were used as the input to another gate. In this case any negative going voltage spikes at the input of the second gate would have to exceed 2.5 V in amplitude (that is, sufficient to lower the voltage level below 2.5 V) before the output voltage would change. The differences between the normal 'low' and 'high' output voltages of a gate and the levels at which the voltage category changes from 'low' to 'high' and vice versa is a measure of the *noise immunity* or *noise margin* of the gate.

COMBINATIONAL LOGIC CIRCUITS

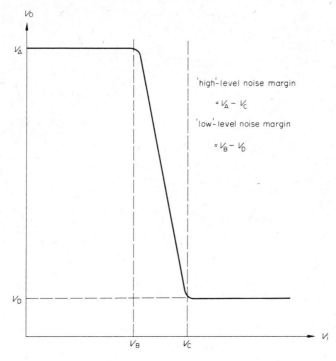

Figure 1.5

1.6 NAND and NOR gates

This is a further case in which examination of the discrete component version of the gate can help illuminate the operation, although in practice, most gates of this type are currently used in integrated circuit form. The circuit used in figure 1.6 has some deficiencies in practice owing to the degree of precision required in specifying resistor values for satisfactory operation. It will however serve to introduce this type of gate.

Consider the circuit shown in figure 1.6. If input B is held at 0 V, the transfer characteristic relating input voltage A to output voltage F will be similar to that shown in figure 1.2 for the NOT gate. Similarly, if input voltage A is held at 0 V then the relationship between input voltage B and output voltage F is as shown in figure 1.2. The behaviour of this gate may be summarised as follows:

If input voltage A or input voltage B is 'high' then the output voltage is 'low'. If input voltage A and input voltage B are both 'high' then the output voltage is 'low'.

An alternative statement of this action would be:

The only situation in which the output voltage F is 'high' is when the voltage levels at inputs A and B are both 'low'.

These statements are cumbersome and the gate operation is described much more elegantly by the truth table shown in table 1.4.

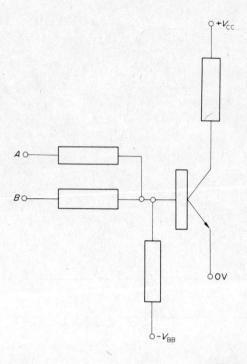

Figure 1.6

Table 1.4

A	B	F
L	L	H
L	H	L
H	L	L
H	H	L

1.7 The NOR operation

Reconstructing the truth table of table 1.4 using **0** and **1** defined by the rules of positive logic (that is, **1** represents a 'high' voltage, **0** represents a 'low' voltage) gives the truth table shown in table 1.5.

COMBINATIONAL LOGIC CIRCUITS

Table 1.5

A	B	F
0	0	1
0	1	0
1	0	0
1	1	0

The operation indicated by this truth table can be described as follows

when A or B is 1 then the output F is not 1

This is the logical NOR operation. It should be noted carefully that the 'or' in the above statement should really be 'and/or' since the output of the gate is 0 when A and B are both 1. The ideas outlined above for a two-input gate are easily extended to include more inputs. The symbol and truth table for a three-input NOR gate are shown in figure 1.7 and table 1.6.

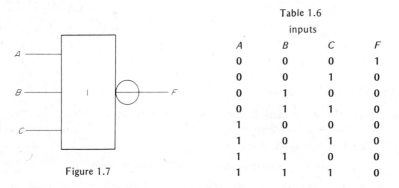

Figure 1.7

Table 1.6

inputs			
A	B	C	F
0	0	0	1
0	0	1	0
0	1	0	0
0	1	1	0
1	0	0	0
1	0	1	0
1	1	0	0
1	1	1	0

The gate shown in figure 1.6 performs the logical NOR operation when positive logic is specified. It is possible to design circuits which will perform the NOR operation with negative logic. It is not necessary for the student to be familiar with the details of such circuits at this stage, merely that he should accept the fact that such circuits are possible.

1.8 The NAND operation

Suppose the truth table shown in table 1.4 is reconstructed according to the rules of negative logic (that is, 'low' voltages are represented by 1, 'high' voltages are represented by 0). The resulting truth table is shown in table 1.7.

Table 1.7

A	B	F
1	1	0
1	0	1
0	1	1
0	0	1

The operation specified by the truth table can be described as follows

> when inputs A and B are both 1 then the output F is not 1

or alternatively

> the only situation in which the output F can be 0 is when A and B are both 1

This is the logical NAND operation. The symbol and truth table for a three-input NAND gate are shown in figure 1.8 and table 1.8 respectively.

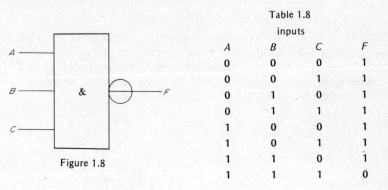

Figure 1.8

Table 1.8

A	B	C	F
0	0	0	1
0	0	1	1
0	1	0	1
0	1	1	1
1	0	0	1
1	0	1	1
1	1	0	1
1	1	1	0

1.9 The NAND–NOR relationship

It has now been demonstrated that the gate shown in figure 1.6 will perform the logical NOR operation if positive logic is specified or the logical NAND operation if negative logic is specified. Hence

$$\text{NOR (positive logic)} \equiv \text{NAND (negative logic)}$$

A similar relationship between the NAND and NOR operations is observed by considering the truth table shown in table 1.9.

Table 1.9

A	B	F
L	L	H
L	H	H
H	L	H
H	H	L

This table is reconstructed using the rules of positive logic in table 1.10 and using negative logic in table 1.11.

Table 1.10

inputs		output
A	B	F
0	0	1
0	1	1
1	0	1
1	1	0

nand

Table 1.11

inputs		output
A	B	F
1	1	0
1	0	0
0	1	0
0	0	1

nor

COMBINATIONAL LOGIC CIRCUITS

It will be seen that table 1.10 describes the logical NAND operation and table 1.11 describes the logical NOR operation. This demonstrates that a circuit which performs the positive logic NAND operation also performs the negative logic NOR operation. Hence

NAND (positive logic) ≡ NOR (negative logic)

1.10 AND and OR gates

Consider a two-input gate which behaves in accordance with the truth table shown in table 1.12

Table 1.12

inputs		output
A	B	F
L	L	L
L	H	L
H	L	L
H	H	H

1.11 The AND operation

Using positive logic the truth table of table 1.12 is converted to that shown in table 1.13.

Table 1.13

inputs		output
A	B	F
0	0	0
0	1	0
1	0	0
1	1	1

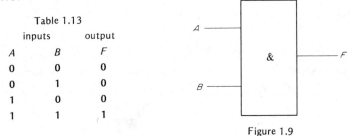

Figure 1.9

The operation defined by this truth table may be described as follows

when input A and input B are both 1, then the output F is 1

This defines the logical AND operation. The symbol for this type of gate is shown in figure 1.9.

1.12 The OR operation

If negative logic is used, the truth table shown in table 1.12 is reconstructed using 1 and 0 in place of L and H to give the table shown in table 1.14.

Table 1.14

inputs		output
A	B	F
1	1	1
1	0	1
0	1	1
0	0	0

The operation defined by this truth table may now be described as follows

when input A or input B is 1 then the output F is 1

This defines the logical OR operation.

The symbol for this type of gate is shown in figure 1.10. The reader should note at this point that the OR operation, as defined above, implicitly includes the AND operation since the truth table in table 1.14 shows that the output of the gate is 1 if either of the two inputs is 1 but the output is also 1, if both of the two inputs are 1.

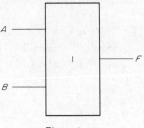

Figure 1.10

1.13 The AND-OR relationships

The gate described by the truth table shown in table 1.12 has been shown to perform the logical AND operation with positive logic and the logical OR operation with negative logic.

$$\text{AND (positive logic)} \equiv \text{OR (negative logic)}$$

A similar situation is found to exist if we examine the truth table shown in table 1.15.

Table 1.15

inputs		output
A	B	F
L	L	L
L	H	H
H	L	H
H	H	H

This converts to the table in table 1.16 for positive logic and to the table in table 1.17 for negative logic.

Table 1.16				Table 1.17		
inputs		output		inputs		output
A	B	F		A	B	F
0	0	0		1	1	1
0	1	1		1	0	0
1	0	1		0	1	0
1	1	1		0	0	0

COMBINATIONAL LOGIC CIRCUITS

It will be seen that table 1.16 defines the logical OR operation and table 1.17 defines the logical AND operation. Hence

$$\text{OR (positive logic)} \equiv \text{AND (negative logic)}$$

Manufacturers of integrated circuits may offer only one type of logical operation in their range of circuits. A very popular range for example, until quite recently offered only positive logic NAND gates with various numbers of inputs. These could of course be operated as negative logic NOR gates if desired.

1.14 Interconnection of gates

Simple gates may be interconnected to perform logical operations which may not be possible with single gates. Consider a two-input NOR gate followed by a NOT gate.

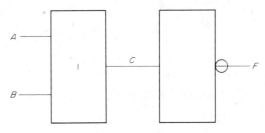

Figure 1.11

The relationship between A, B, C and F is easily shown using the truth table in table 1.18.

Table 1.18

A	B	C	F
0	0	1	0
0	1	0	1
1	0	0	1
1	1	0	1

If the overall system is now considered as consisting of a circuit with two inputs A and B and an output F, it will be seen that the relationship between A, B and F is identical to that for the OR operation shown in table 1.14. This demonstrates the following relationship

$$\text{NOR followed by NOT} \equiv \text{OR}$$

The relationship shown below may be demonstrated in a similar manner

$$\text{OR followed by NOT} \equiv \text{NOR}$$

Similar forms of relationships follow for NAND and NOT gates. The circuit shown in figure 1.12 gives rise to the truth table in table 1.19.

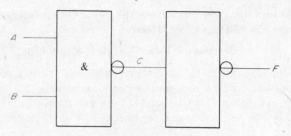

Figure 1.12

Table 1.19

A	B	C	F
0	0	1	0
0	1	1	0
1	0	1	0
1	1	0	1

This shows the following relationship

   NAND followed by NOT ≡ AND

And it follows similarly that

   AND followed by NOT ≡ NAND

A more complicated multi-gate system is shown in figure 1.13.

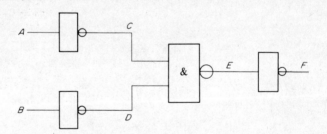

Figure 1.13

The truth table relating A, B, C, D, E and F is shown in table 1.20.

Table 1.20

A	B	C	D	E	F
0	0	1	1	0	1
0	1	1	0	1	0
1	0	0	1	1	0
1	1	0	0	1	0

COMBINATIONAL LOGIC CIRCUITS

If this is considered as a single circuit with inputs A and B and output F, the truth table for the system reduces to table 1.21.

Table 1.21

A	B	F
0	0	1
0	1	0
1	0	0
1	1	0

It will be seen that table 1.21 is equivalent to table 1.5 indicating that the circuit of figure 1.13 performs the NOR operation using NAND and NOT gates only. If the final NOT gate is omitted then the circuit performs the logical OR operation.

In some cases, a NOT gate may be required in a situation which would involve the inclusion of an additional integrated circuit chip in the system. If a spare NAND or NOR gate is available on one of the chips already in the circuit, the NOT operation is easily implemented by holding all but one of the inputs of a NAND gate at **1** or all but one of the inputs of a NOR gate at **0**. A similar function is obtained by connecting together all inputs of a NAND or NOR gate.

Table 1.22

A	B	F
0	1	1
1	1	0

Table 1.23

A	B	F
0	0	1
1	0	0

Table 1.22 indicates that the relationship between the input A and output F in this case is in accordance with the NOT operation. Table 1.23 indicates that the NOT operation is also performed by a two-input NOR gate with one input held at **0**.

Problems

1 Construct a truth table showing the relationship between A, B and F for the system shown in figure 1.14.

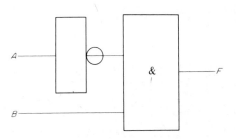

Figure 1.14

2 Show that the system in figure 1.15 performs the logical AND operation. How could it be modified to perform a NAND operation?

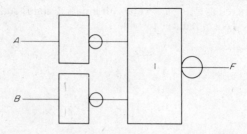

Figure 1.15

3 Figure 1.16 shows the symbol and table 1.24 shows the truth table for a special circuit called an exclusive OR gate. Show that if the two inputs to the gate are preceded by NOT gates, the resulting system still performs the exclusive OR operation.

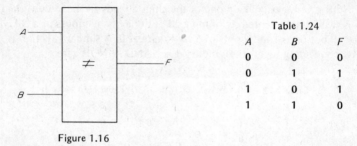

Figure 1.16

Table 1.24

A	B	F
0	0	0
0	1	1
1	0	1
1	1	0

4 Give truth tables for (a) a three-input AND gate; (b) a four-input OR gate.

5 Examine the system shown in figure 1.13, Could a three-input NAND gate be used with NOT gates in the same way to perform like a three-input NOR gate?

2

Boolean Algebra

Having examined the behaviour of some simple gates using truth table techniques, it is now necessary to investigate some of the mathematical techniques available for the analysis and syntheses of multigate logic systems. In 'conventional' mathematical analysis, the use of symbols to represent algebraic variables and the derivation of the various rules by which these variables may be manipulated greatly simplifies the solving of problems. The use of symbols to represent the logic levels at the inputs and outputs of gates and the use of rules by which these quantities may be manipulated similarly assists the understanding and design of logic systems. The variables involved in logic systems are however rather different to ordinary algebraic variables in that each variable has one of only two possible values.

2.1 George Boole

In 1847, the mathematician George Boole devised a system for manipulating items which could have only two possible values. Boole was in fact concerned with the symbolic representation of a logical system first investigated by Aristotle involving statements which were true or false with no intermediate states possible. The work of Boole was a little more than a mathematical curiosity for many years until 1937 when C. E. Shannon recognised that Boole's methods could be used for the analysis of telephone switching networks.

2.2 Boolean variables

Boolean algebra can be used in any situation involving *boolean variables*. These are variables which can exist in one of only two possible states. Boolean variables can identify measurable items such as the voltage levels in logic circuits or more abstract things such as the truth or falsity of a statement. Because boolean variables are so different from ordinary algebraic variables, it is not surprising that the rules for manipulating these variables are very different from the rules of ordinary algebra. The student should attempt to approach the subject with an open mind with no preconceptions as to how boolean variables should behave.

2.3 Boolean inversion

If a boolean variable is denoted by A then the boolean inverse of A is denoted by $\overline{A}$ which is read 'NOT A'. If the possible values of A are 0 and 1, then

$$\overline{1} = 0$$

and

$$\overline{0} = 1$$

This definition of boolean inversion enables the operation of a NOT gate to be described. If the gate has input A and output F then the gate operation is described in symbolic form by

$$F = \overline{A}$$

This means that when

$$A = 1, \quad F = 0$$

and when

$$A = 0, \quad F = 1$$

2.4 The boolean AND operation

If A and B are two boolean variables, the boolean AND operation involving A and B is denoted by $A \cdot B$, that is, 'A AND B'. The operation of an AND gate with inputs, A and B and output F is described by

$$F = A \cdot B$$

that is

F is 1 when A and B are both 1.

2.5 The boolean OR operation

The OR operation involving the two boolean variables A and B is denoted by $A + B$ that is 'A OR B'. The operation of an OR gate with inputs A and B and output F is described in symbolic terms by

$$F = A + B$$

that is

F is 1 when A or B is 1.

This OR operation as before is inclusive of the AND operation.

2.6 The NAND operation

It has been demonstrated that a NAND operation is performed by an AND gate followed by a NOT gate. This implies the following relationship between the inputs A and B and output F for a NAND gate

$$F = \overline{A \cdot B}$$

2.7 The NOR operation

This may be performed by an OR gate followed by a NOT gate and is therefore described by the following relationship

$$F = \overline{A + B}$$

2.8 Boolean expressions and truth tables

The operation of multigate systems may be described by truth tables or by the appropriate boolean expression. It is important that the student should be able to convert from one form to the other easily and quickly.

Consider a multigate system whose operation is described by the boolean relationship

$$F = A \cdot B + \overline{B \cdot C}$$

where A, B, C are the system inputs and F is the output. If we put $A \cdot B = P$ and $\overline{B \cdot C} = Q$ then the operation of the system may be described by the following statement

$$F \text{ is } 1 \text{ when } P \text{ or } Q \text{ is } 1$$

A truth table is constructed by putting down all possible combinations of 1 and 0 for the three inputs A, B, C and forming first P and Q for each set of inputs and finally F as follows:

Table 2.1

A	B	C	$P = A \cdot B$	$B \cdot C$	$Q = \overline{B \cdot C}$	$F = P + Q$
0	0	0	0	0	1	1
0	0	1	0	0	1	1
0	1	0	0	0	1	1
0	1	1	0	1	0	0
1	0	0	0	0	1	1
1	0	1	0	0	1	1
1	1	0	1	0	1	1
1	1	1	1	1	0	1

As a further example, consider a circuit with an output F where F is given by

$$F = A \cdot B \cdot \bar{C} + \overline{(A \cdot \bar{B} \cdot C)}$$

The truth table is constructed in stages as in the previous example.

Table 2.2

A	B	C	$\bar{B}$	$\bar{C}$	$P = A \cdot B \cdot \bar{C}$	$A \cdot \bar{B} \cdot C$	$Q = \overline{A \cdot \bar{B} \cdot C}$	$F = P + Q$
0	0	0	1	1	0	0	1	1
0	0	1	1	0	0	0	1	1
0	1	0	0	1	0	0	1	1
0	1	1	0	0	0	0	1	1
1	0	0	1	1	0	0	1	1
1	0	1	1	0	0	1	0	0
1	1	0	0	1	1	0	1	1
1	1	1	0	0	0	0	1	1

Conversion of the truth table representation of a logical operation to a boolean representation is also quite simple. Consider a logic system with inputs A, B, C and output F whose behaviour is in accordance with the truth table shown in table 2.3.

Table 2.3

A	B	C	F
0	0	0	0
0	0	1	0
0	1	0	1
0	1	1	0
1	0	0	0
1	0	1	0
1	1	0	0
1	1	1	1

From row three of the truth table it will be seen that F must be 1 when A is 0, B is 1 and C is 0. This is stated formally as follows:

F is 1 when A is not 1 and B is 1 and C is not 1

In symbolic form this may be written $F = \bar{A} \cdot B \cdot \bar{C}$. In addition to this, row eight of the truth table also indicates that the following statement is true

F is 1 when A is 1 and B is 1 and C is 1

Hence $F = A \cdot B \cdot C$.

The output F must be 1 therefore when either of the above two statements is true. The complete behaviour of the circuit may be represented in symbolic terms as follows

$$F = \bar{A} \cdot B \cdot \bar{C} + A \cdot B \cdot C$$

As a further example of the technique, consider a system with two inputs A and B and output F whose operation is described by the truth table shown in table 2.4.

Table 2.4

A	B	F
0	0	1
0	1	0
1	0	1
1	1	1

From line one $\quad F = \bar{A} \cdot \bar{B}$
From line three $\quad F = A \cdot \bar{B}$
From line four $\quad F = A \cdot B$

The operation of the circuit is therefore described by the following relationship

$$F = \bar{A} \cdot \bar{B} + A \cdot \bar{B} + A \cdot B$$

2.9 Simplification of boolean expressions

It is often possible to simplify a boolean expression by making use of some of the basic relationships which exist between boolean variables. This can lead to a reduction in the total number of gates required to implement a logical operation,

BOOLEAN ALGEBRA

familiarity with these boolean relationships, particularly De Morgan's theorems is also necessary when converting OR type expressions to AND types and vice versa.

2.10 Summary of boolean relationships

A boolean variable A has two possible values indicated by 0 or 1.

$\bar{A}$ the boolean inverse of A
$\bar{0} = 1$
$\bar{1} = 0$

$A \cdot B$ defines the boolean operation A AND B
$A + B$ defines the boolean operation A OR B
$\overline{A \cdot B}$ defines the boolean operation NAND
$\overline{A + B}$ defines the boolean operation NOR

$$A + A = A$$
$$A + \bar{A} = 1$$
$$A + 0 = A$$
$$A + 1 = 1$$
$$A \cdot A = A$$
$$A \cdot \bar{A} = 0$$
$$A \cdot 0 = 0$$
$$A \cdot 1 = A$$

De Morgan's theorems

(1) $\quad \bar{A} + \bar{B} = \overline{A \cdot B}$
(2) $\quad \bar{A} \cdot \bar{B} = \overline{A + B}$

It follows from these theorems that

$$A \cdot B = \overline{\bar{A} + \bar{B}}$$
$$A + B = \overline{\bar{A} \cdot \bar{B}}$$
$$(A + \bar{A}B) = (A + B)$$
$$A(1 + B) = A$$

Commutativity rule

$$A + B = B + A$$
$$A \cdot B = B \cdot A$$

Associativity rule

$$A + (B + C) = (A + B) + C$$
$$A \cdot (B \cdot C) = (A \cdot B) \cdot C$$

Distributivity rule

$$A + B \cdot C = (A + B) \cdot (A + C)$$
$$A \cdot (B + C) = AB + AC$$

The validity of any of these rules is easily checked by the use of truth tables to indicate the values of the left- and right-hand sides of the equations for all possible sets of input values.

As an example, consider the first distributivity rule statement $A + B \cdot C = (A + B) \cdot (A + C)$. A truth table is formed to find $F_1 = A + B \cdot C$ and $F_2 = (A + B) \cdot (A + C)$ for all possible sets of values for A, B and C.

Table 2.5

A	B	C	$B \cdot C$	$(A+B)$	$(A+C)$	$F_1 = A + B \cdot C$	$F_2 = (A+B) \cdot (A+C)$
0	0	0	0	0	0	0	0
0	0	1	0	0	1	0	0
0	1	0	0	1	0	0	0
0	1	1	1	1	1	1	1
1	0	0	0	1	1	1	1
1	0	1	0	1	1	1	1
1	1	0	0	1	1	1	1
1	1	1	1	1	1	1	1

Examination of the last two columns of table 2.5 indicates that they are identical and hence $F_1 = F_2$ for all A, B, C and demonstrates the validity of the relationship.

At this stage, note the following carefully

$$\overline{A} + \overline{B} \neq \overline{A + B}$$
$$\overline{A} \cdot \overline{B} \neq \overline{A \cdot B}$$

These inequalities may be easily checked by means of a truth table.

The next section demonstrates how application of these rules can greatly assist the design of economical systems to perform simple logical operations.

Problems

1. Without using truth tables prove that $A \cdot (B + \overline{B}) = A$.
2. Use the results of problem 1 to prove the following relationship

$$A \cdot B + A \cdot \overline{B} + C \cdot B + C \cdot \overline{B} = A + C$$

3. Without using truth tables, prove that

$X \cdot Y \cdot Z + X \cdot \overline{Y} \cdot Z + X \cdot Y \cdot \overline{Z} = X \cdot Y + X \cdot Z$
  (Hint: since $A + A = A$ then $X \cdot Y \cdot Z = X \cdot Y \cdot Z + X \cdot Y \cdot Z$)

4. Use truth tables to prove both forms of the De Morgan relationship.
5. Show by means of truth tables that De Morgan's relationships can be extended to three variables.
6. Use De Morgan's relationship to show that $A \cdot (B + C) = \overline{\overline{A} + (\overline{B} \cdot \overline{C})}$.
7. Give a boolean expression to describe the operation of the exclusive OR gate introduced in problem 1.

BOOLEAN ALGEBRA

Bibliography

1. G. Boole. *An Investigation into the Laws of Thoughts*, Dover (1954).
2. C. E. Shannon. 'Symbolic Analysis of Relay and Switching Circuits', *Trans. Am. Inst. elect. Engrs*, **57**, (1938) 713-23.
3. F. J. Hill and G. R. Peterson. *Introduction to Switching Theory and Logical Design*, Wiley (1968).
4. E. Mendelson. *Boolean Algebra and Switching Circuits*, McGraw-Hill (1970).

Handwritten annotations:

$A + B \cdot C = (A+B) \cdot (A+C)$

+ or
· and

$F = A \cdot (B+C) = \bar{A} + (\bar{B} \cdot \bar{C})$

$A \cdot B + A \cdot C = \overline{A+B} \cdot \overline{A+C}$

$\bar{F} = \overline{A \cdot B + A \cdot C} = \overline{\overline{A+B} \cdot \overline{A+C}}$

$= \overline{A \cdot B} \cdot \overline{A \cdot C} = \overline{\overline{A+B}} + \overline{\overline{A+C}}$

$= \overline{A+B} \cdot \overline{A+C} = A \cdot B + A \cdot C$

$A \cdot B \cdot A \cdot C$

$F = \overline{A+B} \cdot \overline{A+C}$

3

Design of Simple Logic Systems

3.1 A simple design example

As an introduction to simple logic design procedures, we shall consider a system which is required to act as a safety interlock to prevent an electric motor running in a situation when damage can result. Such safety interlock systems have been used for many years of course and often use relay circuits. Although the systems may be and often are designed by purely intuitive means, boolean techniques may be applied and will be used in this case.

The problem is stated as follows. A motor is required to run when the 'power on' switch is operated provided that the safety guard is in position and provided that the current taken from the supply does not exceed some predetermined value. We assume that suitable transducers are available which give voltage levels falling into the logical 1 category when the safety guard is in position and when the current taken from the supply is below the safe limit. The system is required to give a logical 1 output when the above conditions are satisfied. This output could then be used to close a relay thus connecting the motor to the power supply.

Let the outputs from the transducers connected to the 'power on' switch, the safety guard and the supply current indicator be described by the boolean variables P, G and C respectively. The required system operation is described by the truth table shown in table 3.1.

Table 3.1

P	G	C	F
0	0	0	0
0	0	1	0
0	1	0	0
0	1	1	0
1	0	0	0
1	0	1	0
1	1	0	0
1	1	1	1

DESIGN OF SIMPLE LOGIC SYSTEMS

It is immediately apparent that table 3.1 simply describes the boolean AND operation. The required system could therefore consist of a single AND gate. Suppose, however, gates are to be used from a range in which only positive logic NOR type operations are available. An obvious step is to use De Morgan's theorem to convert the AND type operation to the equivalent NOR operation. We have

$$\bar{P} + \bar{G} + \bar{C} = \overline{P \cdot G \cdot C}$$

Inverting both sides gives

$$\overline{\bar{P} + \bar{G} + \bar{C}} = \overline{\overline{P \cdot G \cdot C}} = P \cdot G \cdot C$$

This shows that an AND operation on P, G and C is equivalent to a NOR operation on $\bar{P}$, $\bar{G}$ and $\bar{C}$. The required operation is therefore performed using NOR gates only in the system shown in figure 3.1. Note that two-input NOR gates with one input taken to logical **0** have been used to perform the NOT operation as described previously.

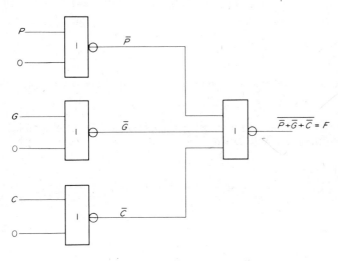

Figure 3.1

3.2 An additional variable

Let us now consider the design problem discussed above but with the additional feature that a maintenance engineer shall have a key which when inserted will enable the motor to run whether or not the safety guard is in position although the 'power on' switch must still be operated and the supply current must be below the safety limit. A device can be incorporated in the system such that insertion of the key will cause a voltage level identified as a logic **1** to be generated. Let the output of the system indicating whether or not the key is in position be denoted by the boolean variable K.

The truth table describing the operation of the system must be modified to take account of the new feature of the safety guard override key. The revised truth table is shown in table 3.2.

Table 3.2

P	G	C	K	F
0	0	0	0	0
0	0	0	1	0
0	0	1	0	0
0	0	1	1	0
0	1	0	0	0
0	1	0	1	0
0	1	1	0	0
0	1	1	1	0
1	0	0	0	0
1	0	0	1	0
1	0	1	0	0
1	0	1	1	1
1	1	0	0	0
1	1	0	1	0
1	1	1	0	1
1	1	1	1	1

Table 3.2 indicates that the symbolic description of the system is as follows

$$F = P \cdot \bar{G} \cdot C \cdot K + P \cdot G \cdot C \cdot \bar{K} + P \cdot G \cdot C \cdot K$$

Implementation of this operation using AND, OR and NOT gates is straightforward and the resulting system is shown in figure 3.2.

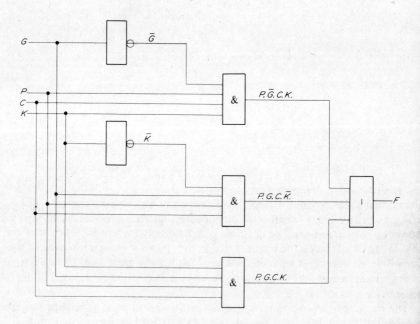

Figure 3.2

DESIGN OF SIMPLE LOGIC SYSTEMS

In the form shown however, the system uses more gates than are necessary. A considerable simplification is possible by applying some of the rules of boolean algebra.

$$F = P \cdot \bar{G} \cdot C \cdot K + P \cdot G \cdot C \cdot \bar{K} + P \cdot G \cdot C \cdot K$$
$$= P \cdot \bar{G} \cdot C \cdot K + P \cdot G \cdot C \cdot (\bar{K} + K)$$
$$= P \cdot \bar{G} \cdot C \cdot K + P \cdot G \cdot C$$
$$= P \cdot C \cdot (\bar{G} \cdot K + G)$$
$$= P \cdot C \cdot (K + G)$$

The resulting much simplified system is shown in figure 3.3.

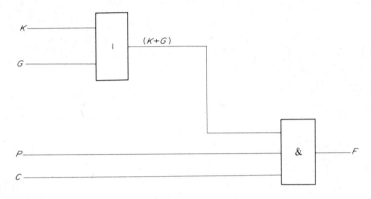

Figure 3.3

Suppose now that we are again required to perform the same operation using positive logic NOR gates only. De Morgan's theorem provides the required link between the AND operations in the original expression and the required NOR form.

From the above analysis, the required operation is given by

$$F = P \cdot C \cdot (K + G)$$
$$= P \cdot C \cdot X \quad \text{where} \quad X = (K + G)$$
$$= \overline{\bar{P} + \bar{C} + \bar{X}}$$
$$= \overline{\bar{P} + \bar{C} + \overline{(K + G)}}$$

Since the final expression involves no AND type operations, it may be implemented using NOR gates only. The system is shown in figure 3.4.

3.3 A second design example

A second simple design example will now be considered. Suppose a system is required with four inputs A, B, C, D whose output F is in accordance with the truth table shown in table 3.3.

Examination of the truth table indicates that the system operation may be described in boolean terms by

$$F = A \cdot \bar{B} \cdot \bar{C} \cdot D + \bar{A} \cdot \bar{B} \cdot C \cdot D + A \cdot B \cdot C \cdot \bar{D}$$

In designing this system it will be assumed that a range of gates is to be used in which only positive logic NAND operations are available. Inspection of the above

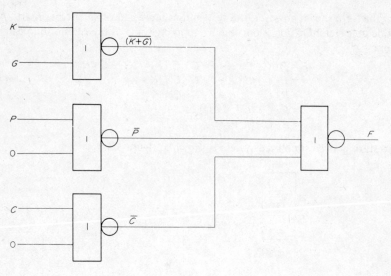

Figure 3.4

Table 3.3

A	B	C	D	F
0	0	0	0	0
0	0	0	1	0
0	0	1	0	0
0	0	1	1	1
0	1	0	0	0
0	1	0	1	0
0	1	1	0	0
0	1	1	1	0
1	0	0	0	0
1	0	0	1	1
1	0	1	0	0
1	0	1	1	0
1	1	0	0	0
1	1	0	1	0
1	1	1	0	1
1	1	1	1	0

boolean expression indicates that little in the way of factorisation is possible but as an initial step we have

$$F = A \cdot \bar{B} \cdot \bar{C} \cdot D + \bar{A} \cdot \bar{B} \cdot C \cdot D + A \cdot B \cdot C \cdot \bar{D}$$
$$= A \cdot (\bar{B} \cdot \bar{C} \cdot D + B \cdot C \cdot \bar{D}) + \bar{A} \cdot \bar{B} \cdot C \cdot D \qquad (3.1)$$

The next stage is to eliminate the OR type operations from the above expression.

DESIGN OF SIMPLE LOGIC SYSTEMS

This is done in stages using De Morgan's theorem.
We have

$$(\bar{B} \cdot \bar{C} \cdot D) + (B \cdot C \cdot \bar{D}) = \overline{(\bar{B} \cdot \bar{C} \cdot D) \cdot (B \cdot C \cdot \bar{D})}$$
$$= \overline{X \cdot Y}$$

where   (3.2)

$$X = \overline{(\bar{B} \cdot \bar{C} \cdot D)}$$
$$Y = \overline{(B \cdot C \cdot \bar{D})}$$

From (3.1) and (3.2)

$$F = A \cdot \overline{(X \cdot Y)} + \bar{A} \cdot \bar{B} \cdot C \cdot D$$
$$= A \cdot \overline{(X \cdot Y)} + Z$$

where

$$Z = \bar{A} \cdot \bar{B} \cdot C \cdot D$$

$$\therefore \quad F = \overline{\overline{A \cdot \overline{(X \cdot Y)}} \cdot \bar{Z}} \quad (3.3)$$

Design of the system can now proceed in stages. Subsections of the circuit are designed first to generate X, Y and Z. As an initial step however, we note that the inverses of all input quantities will be required at some stage and the first subsection of the system is therefore as shown in figure 3.5.

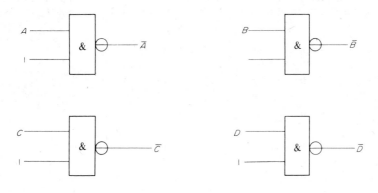

Figure 3.5

Systems are now designed to generate X, Y and Z.

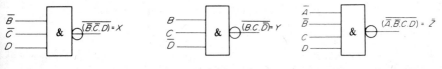

Figure 3.6

The final stage is to design a system which will generate F from inputs A, X, Y and Z. This system is shown in figure 3.7.

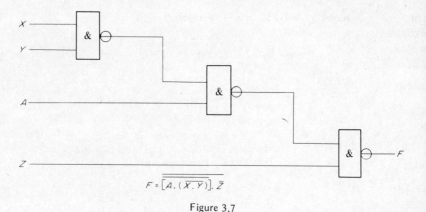

$$F = \overline{[A \cdot (\overline{X \cdot Y})] \cdot \overline{Z}}$$

Figure 3.7

3.4 An equivalence system

As a further demonstration of the design procedure for combinational logic systems, consider the design for a circuit with four inputs A, B, X and Y and two outputs F and G. F is required to be 1 when $A = X$ and $B = Y$. G is required to be 1 when $A \neq X$ and $B \neq Y$. This is an example of a system which can easily be designed intuitively when a little experience has been gained.

In this case, however, we shall carry through the formal design procedure. The first stage is to construct the truth table for the system. This is shown in table 3.4.

Table 3.4

A	B	X	Y	F	G
0	0	0	0	1	0
0	0	0	1	0	0
0	0	1	0	0	0
0	0	1	1	0	1
0	1	0	0	0	0
0	1	0	1	1	0
0	1	1	0	0	1
0	1	1	1	0	0
1	0	0	0	0	0
1	0	0	1	0	1
1	0	1	0	1	0
1	0	1	1	0	0
1	1	0	0	0	1
1	1	0	1	0	0
1	1	1	0	0	0
1	1	1	1	1	0

From the truth table, the system operation can be described in boolean terms by the following two relationships

$$F = \bar{A} \cdot \bar{B} \cdot \bar{X} \cdot \bar{Y} + \bar{A} \cdot B \cdot \bar{X} \cdot Y + A \cdot \bar{B} \cdot X \cdot \bar{Y} + A \cdot B \cdot X \cdot Y \qquad (3.4)$$

$$G = \bar{A} \cdot \bar{B} \cdot X \cdot Y + \bar{A} \cdot B \cdot X \cdot \bar{Y} + A \cdot \bar{B} \cdot \bar{X} \cdot Y + A \cdot B \cdot \bar{X} \cdot \bar{Y} \qquad (3.5)$$

DESIGN OF SIMPLE LOGIC SYSTEMS

We shall again assume that only NAND gates are available to implement the desired operation. The normal procedure of simplification followed by application of De Morgan's theorem to eliminate OR type operations is carried out from (3.4).

$$F = \bar{A} \cdot \bar{X} \cdot (\bar{B} \cdot \bar{Y} + B \cdot Y) + A \cdot X \cdot (\bar{B} \cdot \bar{Y} + B \cdot Y)$$
$$= (\bar{A} \cdot \bar{X} + A \cdot X) \cdot (\bar{B} \cdot \bar{Y} + B \cdot Y)$$

Applying De Morgan's theorem we have

$$(\bar{A} \cdot \bar{X} + A \cdot X) = \overline{\overline{\bar{A} \cdot \bar{X}} \cdot \overline{A \cdot X}}$$

and

$$(\bar{B} \cdot \bar{Y} + B \cdot Y) = \overline{\overline{\bar{B} \cdot \bar{Y}} \cdot \overline{B \cdot Y}}$$

$$\therefore F = \overline{\overline{\bar{A} \cdot \bar{X}} \cdot \overline{A \cdot X}} \cdot \overline{\overline{\bar{B} \cdot \bar{Y}} \cdot \overline{B \cdot Y}} \qquad (3.5)$$

A similar analysis gives

$$G = \overline{\overline{\bar{A} \cdot X} \cdot \overline{A \cdot \bar{X}}} \cdot \overline{\overline{\bar{B} \cdot Y} \cdot \overline{B \cdot \bar{Y}}} \qquad (3.6)$$

As a first step in constructing the system, it is noted that inverses of all inputs are required and these are generated in the usual manner using two input NAND gates with one input held at logical 1. The system is then built up stage by stage as described in the previous examples.

The systems given in this section are not necessarily the most economical and it is quite possible that the number of gates used could be reduced. The problem of designing the most economical or 'minimal' systems is however considerably more complicated than merely designing a simple system which performs the required operation. This is discussed further in the next chapter.

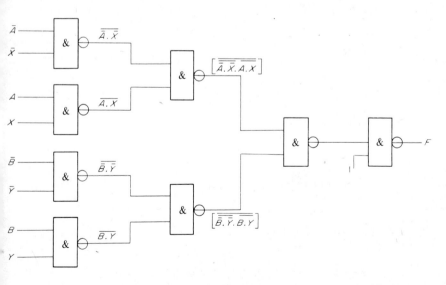

Figure 3.8

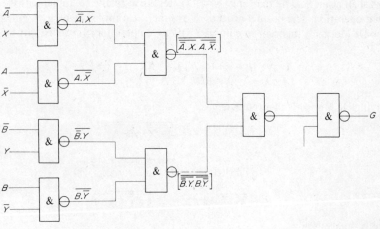

Figure 3.9

Problems

1 Design a logic system using NOR gates only, which performs in accordance with the truth table given in table 3.5.

Table 3.5

inputs			output
A	B	C	F
0	0	0	1
0	0	1	0
0	1	0	0
0	1	1	0
1	0	0	1
1	0	1	0
1	1	0	0
1	1	1	0

2 A control panel has three lamps A, B, C and two switches X and Y. The following situations represent alarm conditions:

a All lamps lit simultaneously.
b Lamps A and B lit when either X or Y is in the ON position.
c Lamp C lit when Y is in the ON position.

Design logic systems to indicate when an alarm condition exists.

(i) Using NAND gates only.
(ii) Using NOR gates only.

3 Give a truth table for the logic system shown in figure 3.10. Describe the operation in boolean terms.

DESIGN OF SIMPLE LOGIC SYSTEMS

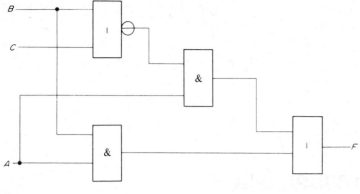

Figure 3.10

4 How would the truth table for the system shown in figure 3.10 be modified if the NOR gate developed a fault such that the output became permanently 1.

5 The system shown in figure 3.10 is to be redesigned using a range of gates in which only NAND operations are available. Show a suitably modified system.

6 A logic system has three inputs A, B, C. The output F is required to be 1 if the total number of 1's at the inputs is an even number (zero is regarded as an even number). Design the system using not more than the following complement of gates:

6 two-input AND gates
3 two-input OR gates
3 NOT gates.

4
Karnaugh Maps

Karnaugh maps together with boolean algebra provide the logic system designer with the two major 'tools of the trade'. Map techniques have certain advantages when looking for relationships which enable the total number of gates required to implement a given logical operation to be minimised. Although traditionally the subject is often approached by an initial study of Venn diagrams, the treatment here will concentrate on Karnaugh maps as modified truth tables. The student will by now be familiar with truth tables as listings of the various possible combinations of input states to a circuit together with the corresponding output states. As an example, we shall take the truth table shown in table 4.1, representing the following boolean relationship

$$F = A \cdot B \cdot C \cdot D + A \cdot \bar{B} \cdot C \cdot D + A \cdot B \cdot C \cdot \bar{D} + \bar{A} \cdot B \cdot C \cdot \bar{D}$$

Table 4.1

A	B	C	D	F
0	0	0	0	0
0	0	0	1	0
0	0	1	0	0
0	0	1	1	0
0	1	0	0	0
0	1	0	1	0
0	1	1	0	1
0	1	1	1	0
1	0	0	0	0
1	0	0	1	0
1	0	1	0	0
1	0	1	1	1
1	1	0	0	0
1	1	0	1	0
1	1	1	0	1
1	1	1	1	1

KARNAUGH MAPS

The information displayed by the truth table may be presented in a slightly different form. The table has sixteen rows, each row corresponding to one of the sixteen possible combinations of 0 and 1 at the input. A matrix of sixteen boxes or cells can therefore be formed as shown in figure 4.1, each cell corresponding to one row of the truth table with the corresponding value of F.

Figure 4.1

	0000 F=0	0001 F=0	0011 F=0	0010 F=0
row 1	0000 F=0	0001 F=0	0011 F=0	0010 F=0
row 2	0100 F=0	0101 F=0	0111 F=0	0110 F=1
row 3	1100 F=0	1101 F=0	1111 F=1	1110 F=1
row 4	1000 F=0	1001 F=0	1011 F=1	1010 F=0

The order in which the various input combinations of 1 and 0 are placed in the cells is important. It will be observed that no box contains a combination of 0's and 1's which differs by more than one digit from the combination in an adjacent box. Take for example the cell containing 0111 together with the four adjacent cells.

Figure 4.2

	1111	
0110	1110	1010
	1100	

This relationship obviously does not extend to diagonally adjacent cells. The reason for this particular arrangement will become apparent later.

Further examination of figure 4.1 shows also that all input combinations in rows 3 and 4 have $A = 1$ whereas all remaining cells have $A = 0$. Columns 2 and 3 contain

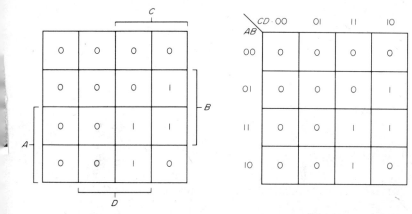

Figure 4.3   Figure 4.4

all input combinations for which $B = 1$ and columns 3 and 4 contain input combinations for which $C = 1$. These facts enable the Karnaugh map to be presented in the form shown in figure 4.3. Cells corresponding to $F = 1$ are shown containing 1.

Some workers prefer the alternative method of identifying the cells in a Karnaugh map which is shown in figure 4.4. This method may be understood by noting that in all cells in column 1 of the map, $C = 0$ and $D = 0$. In column 2, all cells have $C = 0$ and $D = 1$ while in columns 3 and 4, cells have $C = 1, D = 1$ and $C = 1, D = 0$ respectively. A column of the map can therefore be identified by the values of C and D in the column as has been done in figure 4.4. The rows of the map can obviously be identified in a similar way. The map labelled as shown in figure 4.3 is referred to as the Veitch form or occasionally the Venn form whereas the map labelled as in figure 4.4 is referred to as the Karnaugh form. The particular form used is a matter of personal preference. The Karnaugh form probably assists in the construction of the map and labelling the cells although the author is of the opinion that the techniques described later for identifying logic functions from the map are understood more easily using the Veitch form. For this reason, all future maps used for deriving logic functions in the text will be labelled in the Veitch manner. The reader can easily modify the maps to the Karnaugh form if he so desires.

The alert reader may well have observed by now that the arrangement of cells shown in the map described above is not the only one which meets the essential requirement of only one digit difference between labels of adjacent cells. If we examine the Karnaugh form of the labelling in figure 4.4 it is apparent that we could choose column 1 as the column for which we have $A = 0, B = 0$ rather than $C = 0, D = 0$. Row 1 could then have $C = 0, D = 0$ instead of $A = 0, B = 0$. We have thus interchanged the values of A, B and C, D in the rows and columns. The resulting map is shown in figure 4.5. For reasons of uniformity, the cell arrangement shown in figure 4.3 will normally be used but any other arrangement which meets the requirement of one digit difference between adjacent cell labels is equally satisfactory.

CD \ AB	00	01	11	10
00	0000	0100	1100	1000
01	0001	0101	1101	1001
11	0011	0111	1111	1011
10	0010	0110	1110	1010

Figure 4.5

4.1 Looping two cells

Consider now the adjacent cells in rows 3 and 4 of column 3. It will be remembered that the complete map displays the following boolean relationship.

$$F = A \cdot B \cdot C \cdot D + A \cdot \bar{B} \cdot C \cdot D + A \cdot B \cdot C \cdot \bar{D} + \bar{A} \cdot B \cdot C \cdot \bar{D}$$

KARNAUGH MAPS

The two cells under discussion correspond to the terms $A \cdot B \cdot C \cdot D$ and $A \cdot \bar{B} \cdot C \cdot D$ in the above expression, as is seen from figure 4.1. Now, it is possible to simplify the expression for F since we have

$$A \cdot B \cdot C \cdot D + A \cdot \bar{B} \cdot C \cdot D = A \cdot C \cdot D \cdot (B + \bar{B})$$
$$= A \cdot C \cdot D$$

The expression for F therefore reduces to

$$F = A \cdot C \cdot D + A \cdot B \cdot C \cdot \bar{D} + \bar{A} \cdot B \cdot C \cdot \bar{D}$$

A little thought will indicate that possible simplifications of this type will always be associated with two terms represented by adjacent cells containing 1 in the Karnaugh map. This is because adjacent cells correspond to combinations of inputs which differ by one digit only. For the two cells discussed above it will be seen by reference to figure 4.1 that only the second digit is different, corresponding to $B = 1$ in one cell and $B = 0$ in the other.

If we now consider the two adjacent cells in rows 2 and 3 of column 4, the above simplification procedure can again be applied. These cells correspond to the terms $A \cdot B \cdot C \cdot \bar{D}$ and $\bar{A} \cdot B \cdot C \cdot \bar{D}$ in the original expression for F.

$$\bar{A} \cdot B \cdot C \cdot \bar{D} + A \cdot B \cdot C \cdot \bar{D} = B \cdot C \cdot \bar{D}(A + \bar{A})$$
$$= B \cdot C \cdot \bar{D}$$

The expression for F is now reduced to

$$F = A \cdot C \cdot D + B \cdot C \cdot \bar{D}$$

The use of Karnaugh maps therefore provides a method for identifying terms in a logical expression which can be grouped together to give an overall simplification. The advantage of map methods is that the very considerable ability of the human brain to recognise patterns can be brought to bear on the problem of simplifying a logical expression. Pairs of cells which identify possible simplifications are normally shown looped together.*

4.2 The Karnaugh map technique

The formal technique for using Karnaugh maps can now be presented. The map is constructed as described previously and adjacent cells containing 1 are looped together as shown in figure 4.6.

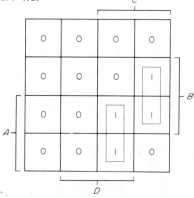

Figure 4.6

If two cells are looped, the corresponding terms in the boolean expression can be combined into a single term *omitting the variable* which is not equal in the two terms. Figure 4.7 and 4.8 show the map technique applied to simplify two other boolean expressions. Figure 4.7 shows the simplification of a three variable expression.

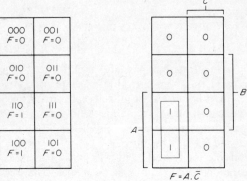

Figure 4.7

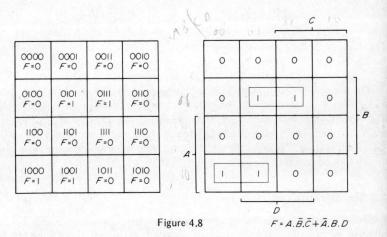

Figure 4.8

4.3 Looping larger groups of cells

Figure 4.9 shows a Karnaugh map representing the boolean relationship

$$F = A \cdot B \cdot C \cdot D + A \cdot \bar{B} \cdot \bar{C} \cdot D + A \cdot B \cdot \bar{C} \cdot D + A \cdot \bar{B} \cdot C \cdot D$$

As described above, the first and last terms on the right-hand side can be combined to give $A \cdot C \cdot D$ and the second pair to give $A \cdot C \cdot D$, corresponding to the looping of cells shown in figure 4.9. This simplifies the original expression to

$$F = A \cdot D \cdot (\bar{C} + C) = A \cdot D$$

Examination of the simplified expression for F indicates immediately that further simplification is possible since

$$F = A \cdot \bar{C} \cdot D + A \cdot C \cdot D$$

KARNAUGH MAPS

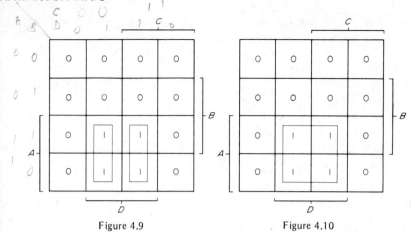

Figure 4.9   Figure 4.10

On the Karnaugh map, this simplification is shown in figure 4.10 by looping together all four adjacent cells containing **1** to form a single group.

The rule for implementing the simplification procedure is similar to that used when looping two cells only. If four cells are looped, four terms in the original boolean expression can be combined into a single term eliminating two variables. The variables which are eliminated are those which are not equal in the four cells. An easy way of identifying which variables are eliminated is to note that if a loop crosses the boundary of a defined *area* of the map then the variable associated with that area is eliminated from the term. For the expression associated with figure 4.10, we note that the loop of four cells cuts across the boundary of the B and C areas of the map indicating that these variables disappear in the resulting term, which is therefore $A \cdot D$. In the above simplification this means that B and C are eliminated. This technique can fairly obviously be extended by looping a group of eight cells. The procedure is similar to that previously described, with a total of three terms eliminated from the boolean expression described by the map.

4.4 Further cell looping techniques

Returning now to figure 4.1, it will be recalled that an important feature of this map is that no cell contains a combination of **0**'s and **1**'s which differs by more than one digit from the combination in an adjacent cell. It is this property which makes it possible to simplify the logical expression associated with the map by looping cells. A further examination of the map reveals, however, that the contents of the cell in row 1, column 1 also differs by one digit only from the contents of the cell in row 1, column 4. This means that if these two cells correspond to an input combination which is required to give a **1** output, then the cell looping technique can again be used to simplify the associated boolean expression. This is also true for all pairs of cells in columns 1 and 4 of any row. Figures 4.11 and 4.12 demonstrate this. As before, each loop contributes a term to the boolean expression. Each term only contains a given variable if the loop does not cut across the area associated with that variable. A variable appearing in a term is inverted if the loop is completely outside the area associated with the variable. In figure 4.11 therefore, since only one loop is present, the function contains only one term. In this term C will be absent since the loop cuts across the boundary of the C area.

Since the loop lies completely outside the *A*, *B* and *D* areas, each of these variables appears inverted in the term, which is therefore $\bar{A} \cdot \bar{B} \cdot \bar{D}$.

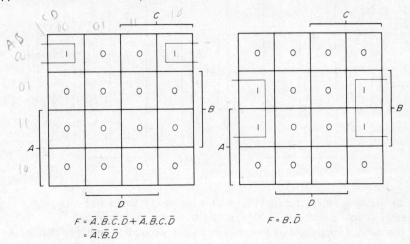

$F = \bar{A}.\bar{B}.\bar{C}.\bar{D} + \bar{A}.\bar{B}.C.\bar{D}$
$\quad = \bar{A}.\bar{B}.\bar{D}$

Figure 4.11

$F = B.\bar{D}$

Figure 4.12

This technique for looping cells which are not apparently adjacent is capable of further expansion since pairs of cells in any column of rows 1 and 4 also have the property that as in figure 4.1 the contents of these cells differ by one digit only. Figure 4.13 demonstrates how cells in rows 1 and 4 are looped.

As a final development of the technique, figure 4.14 demonstrates that four corner cells of a map can be looped.

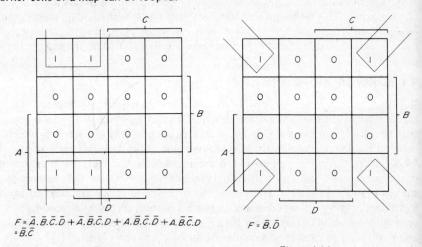

$F = \bar{A}.\bar{B}.\bar{C}.\bar{D} + \bar{A}.\bar{B}.C.D + A.\bar{B}.\bar{C}.\bar{D} + A.\bar{B}.C.D$
$\quad = \bar{B}.\bar{C}$

Figure 4.13

$F = \bar{B}.\bar{D}$

Figure 4.14

It should be noted that when looping cells, the main aims should be to generate loops which are as large as possible and to include as many cells as possible in at least one loop. Once a cell is included in a loop, it is not necessary to attempt to

KARNAUGH MAPS

include that cell in another loop unless this is the only way to complete that loop. Loops must of course be of two, four, eight or sixteen cells. Loops containing any other number of cells are not permissible.

4.5 Looping cells containing 0's

If a particular Karnaugh map represents a boolean function F, a map to represent the function $\bar{F}$ is easily constructed by changing all 1's for 0's and 0's for 1's in the original map. Cells containing 1's in this second map could then be looped to give a simplified expression for $\bar{F}$. Rather than go to the trouble of constructing a new map, it is equally simple to loop cells containing 0's in the map representing F. Since F is easily obtained from $\bar{F}$ by inversion looping 0's represents an alternative to looping 1's to produce a simplified expression for F or $\bar{F}$. Figure 4.15 shows a map with 1's looped and figure 4.16 shows the same map with 0's looped.

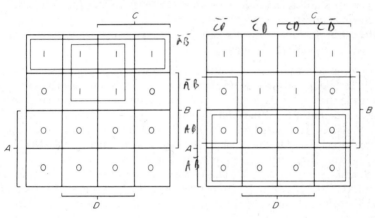

Figure 4.15   Figure 4.16

From figure 4.15, we have

$$F = \bar{A} \cdot \bar{B} + \bar{A} \cdot D$$

Figure 4.16 gives

$$\bar{F} = A + B \cdot \bar{D}$$

or $F = \overline{A + B \cdot \bar{D}}$

It is a fairly simple matter to demonstrate using boolean algebra that the two expressions for F are in fact equivalent. As will be seen in the next section, the type of logic gates available determines whether looping 1's or 0's gives the most convenient simplified expression.

4.6 Karnaugh maps for NAND-NOR logic

In a practical situation, the logic designer may find himself restricted by the fact that the particular range of integrated circuits which he is using contains only a limited range of gate types. It is not uncommon for example to be restricted to only NAND and NOT gates or only NOR and NOT gates. The Karnaugh map methods discussed so far result initially in logical expressions containing a mixture of AND, OR and NOT operations.

De Morgan's theorem can of course be used to eliminate the unwanted operation as has been described in chapter 3. With a little ingenuity however, a logical expression can be obtained from the Karnaugh map which may be implemented directly using a restricted range of gates.

Suppose we wish to use NAND gates only to implement the logical function described by the Karnaugh map shown in figure 4.13. In boolean terms this is

$$F = \bar{A} \cdot \bar{B} + \bar{A} \cdot D$$

Applying De Morgan's theorem, we have

$$F = \overline{(\overline{\bar{A} \cdot \bar{B}}) \cdot (\overline{\bar{A} \cdot D})}$$

All operations in this expression are NAND or NOT. The function is easily implemented directly as shown in figure 4.17.

The use of Karnaugh maps for designing systems using NAND and NOT gates only may be formalised as follows.

(1) Loop cells containing 1's as described in sections 4.2, 4.3 and 4.4.

(2) Each loop corresponds to a term in a boolean expression. Variables appear inverted in the term if the loop is completely outside the area associated with the variable and uninverted if the loop is completely inside the area associated with the term. Variables are omitted from the term if the loop cuts across the boundary of the area associated with the variable. The variables are related by NAND operations to form terms. Thus figure 4.15 represents a function with two terms which are $(\overline{\bar{A} \cdot \bar{B}})$ and $(\overline{\bar{A} \cdot D})$. In cases of loops of eight cells in a sixteen cell map, four cells in an eight cell map or two cells in a four cell map, the loop is identified by a single variable. If such a loop is identified by a single variable X, in the procedure described here, instead of two or more variables linked by a NAND operation, the single variable is inverted to form the term. A loop which covers the entire X area of a map therefore provides a term $\bar{X}$ whereas if the loop is completely outside the X area, it contributes a term X.

(3) The terms thus formed are linked by a NAND operation. Figure 4.15 therefore represents the expression

$$F = \overline{(\overline{\bar{A} \cdot \bar{B}}) \cdot (\overline{\bar{A} \cdot D})}$$

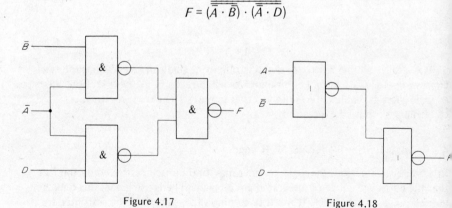

Figure 4.17 Figure 4.18

KARNAUGH MAPS

To generate the NOR form of a boolean expression, it is often more convenient to use the map with 0's looped. From figure 4.16 we have

$$\bar{F} = A + B \cdot \bar{D}$$

Applying De Morgan's theorem to $B \cdot \bar{D}$

$$\bar{F} = A + (\overline{\bar{B} + D})$$

or

$$F = \overline{A + (\overline{\bar{B} + D})}$$

which involves NOR operations only and is implemented directly as shown in figure 4.18. The procedure for generating the NOR form of the function from the Karnaugh map may be formalised as follows.

(1) Loop cells containing 0's as described in section 4.5.
(2) Each loop corresponds to a term. The term in general consists of variables linked by a NOR operation. If the loop is completely within the area associated with a particular variable, the variable appears inverted in the term. If the loop is completely outside the area associated with the variable, the variable appears uninverted. If the loop cuts the boundary of an area associated with a particular variable, that variable does not appear. In the special case of a loop which completely contains the area associated with one variable, that variable appears uninverted as a term. This shown in figure 4.15 which includes a loop which contains the A area of the map. A therefore appears as a term. Similarly, a loop completely outside the A area would contribute the term $\bar{A}$.

From this, the terms arising from figure 4.16 are A and $(\bar{B} + D)$.

(3) The terms are related by NOR operations. The complete expression for the relationship described by figure 4.16 is therefore

$$F = \overline{A + (\overline{\bar{B} + D})}$$

For either the NAND or NOR implementation scheme, if a single loop only is present, the linking operation of step (3) is replaced by an inversion.

4.7 Design example

Consider the design procedure for a system with four inputs A, B, C and D. The logic levels at A and B are assumed to represent a binary number N_1 with A as the most significant digit. The logic levels at C and D represent the number N_2 with C as the most significant digit. If N_1 is greater than N_2, the output F is required to be 1 and for all other conditions the output must be 0. This means that for the condition $A = 0$, $B = 1$, $C = 1$ and $D = 1$, F should be 0 since $N_1 = 01$ and $N_2 = 11$, whereas for the condition $A = 1$, $B = 0$, $C = 0$ and $D = 1$, F should be 1 since $N_1 = 10$ and $N_2 = 01$ and so on.

As a first step, the truth table shown in table 4.2 is prepared. The Karnaugh map is then easily constructed using the methods previously described. Cell looping is shown in three separate stages in figure 4.19 to avoid undue complication of the diagram.

Table 4.2

A	B	C	D	F
0	0	0	0	0
0	0	0	1	0
0	0	1	0	0
0	0	1	1	0
0	1	0	0	1
0	1	0	1	0
0	1	1	0	0
0	1	1	1	0
1	0	0	0	1
1	0	0	1	1
1	0	1	0	0
1	0	1	1	0
1	1	0	0	1
1	1	0	1	1
1	1	1	0	1
1	1	1	1	0

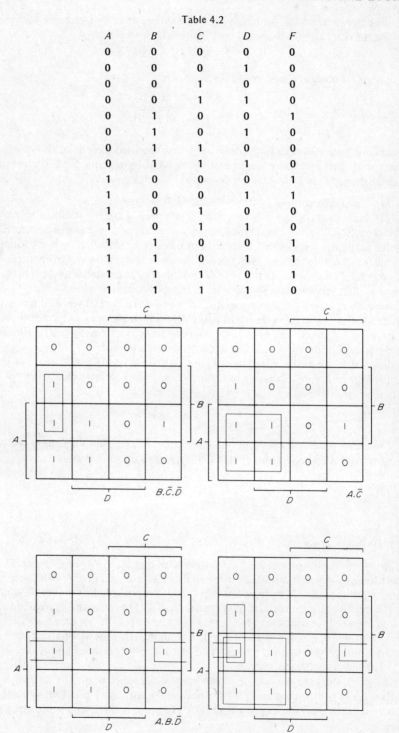

Figure 4.19

KARNAUGH MAPS

From figure 4.19 it follows that the required boolean expression describing the behaviour of the system is

$$F = B \cdot \bar{C} \cdot \bar{D} + A \cdot \bar{C} + A \cdot B \cdot \bar{D}$$

The system design is simply developed from this expression.

4.8 A second example

Suppose the system specification given in the first example is changed and it is required that F should be 1 for N_1 greater than N_2 or $N_1 = N_2$ but be 0 for all other conditions. The modified truth table is shown in table 4.3.

Table 4.3

A	B	C	D	F
0	0	0	0	1
0	0	0	1	0
0	0	1	0	0
0	0	1	1	0
0	1	0	0	1
0	1	0	1	1
0	1	1	0	0
0	1	1	1	0
1	0	0	0	1
1	0	0	1	1
1	0	1	0	1
1	0	1	1	0
1	1	0	0	1
1	1	0	1	1
1	1	1	0	1
1	1	1	1	1

Figure 4.20 shows the various cell loopings.

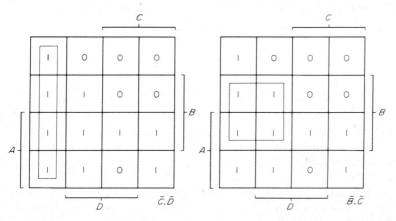

Figure 4.20   [Figure 4.20 continued over

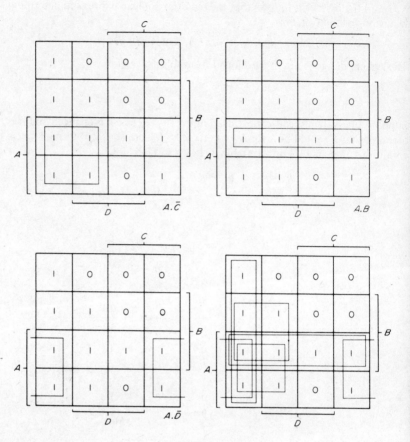

Figure 4.20 continued

The boolean expression describing the circuit operation now becomes

$$F = \bar{C} \cdot \bar{D} + B \cdot \bar{C} + A \cdot B + A \cdot \bar{D} + A \cdot \bar{C}$$

4.9 'Don't care' situations

We now suppose that in the previous problem, the specification for the system is again changed this time such that the value of F for the condition $N_1 = N_2$ is irrelevant, that is it may be 0 or 1. This causes the truth table to be modified as shown in table 4.4.

In the Karnaugh maps shown in figure 4.21, the 'don't care' cells containing a star may be assumed to contain 0 or 1 to suit one's own convenience. This leads to some simplification of the logical expression describing the circuit operation.

KARNAUGH MAPS

Table 4.4

A	B	C	D	F
0	0	0	0	*
0	0	0	1	0
0	0	1	0	0
0	0	1	1	0
0	1	0	0	1
0	1	0	1	*
0	1	1	0	0
0	1	1	1	0
1	0	0	0	1
1	0	0	1	1
1	0	1	0	*
1	0	1	1	0
1	1	0	0	1
1	1	0	1	1
1	1	1	0	1
1	1	1	1	*

* indicates a 'don't care' situation

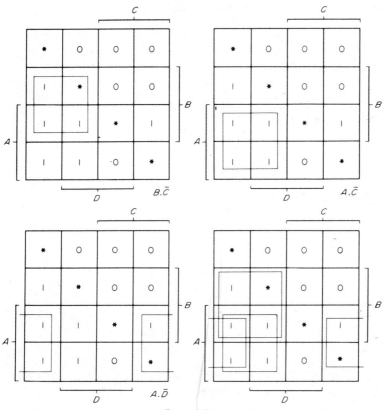

Figure 4.21

By assuming that some 'don't care' cells contain 1 and others contain 0, an economical decision can be made resulting in the expression

$$F = B \cdot \bar{C} + A \cdot \bar{C} + A \cdot \bar{D}$$

4.10 Minimisation with functions of more than four variables

The Karnaugh map technique can be extended to handle functions of five, six or more variables by using maps consisting of arrays of cells in three or more dimensions. The method does, however, lose much of its elegance and simplicity when extended beyond functions of four variables. Various tabular methods have been developed which are more appropriate when attempting to simplify these more complex functions. Readers wishing to pursue this topic further are referred to 3 to 6 of the bibliography.

Problems

1. The truth table shown in table 4.5 expresses a relationship between the four inputs A, B, C, D to a system and the three outputs F, G and H. Use Karnaugh maps to derive appropriate simplified expressions for F, G and H.

Table 4.5

A	B	C	D	F	G	H
0	0	0	0	0	0	1
0	0	0	1	0	0	0
0	0	1	0	0	1	1
0	0	1	1	0	0	1
0	1	0	0	0	0	0
0	1	0	1	0	1	0
0	1	1	0	0	1	1
0	1	1	1	0	1	1
1	0	0	0	1	0	1
1	0	0	1	1	0	0
1	0	1	0	1	1	1
1	0	1	1	1	0	0
1	1	0	0	0	0	0
1	1	0	1	1	0	1
1	1	1	0	0	1	0
1	1	1	1	1	0	0

2. A logic system has four inputs A, B, C and D. The logic levels of the inputs represent a four-digit binary number with A as the most significant digit. The output F of the system is required to be 1 if the number at the input is less than or equal to eight. If the number at the input is above eleven then the value of the output can be 0 or 1.

   With the help of a Karnaugh map, show that the operation of the system is described by the boolean relationship

$$F = \bar{A} + \bar{C} \cdot \bar{D}$$

3 The inputs to a logic system are as described in problem 2. Use Karnaugh maps to design circuits whose output is **1**.

   (a) when the number at the input is a prime number.
   (b) when the number at the output is divisible by four.

4 By means of a Karnaugh map, show that the term $B \cdot D$ is redundant in the following expressions.

$$F = \bar{A} \cdot D + A \cdot B + B \cdot D$$
$$F = B \cdot C + \bar{C} \cdot D + B \cdot D$$

5 Cells containing **0** can be looped in a Karnaugh map to give an expression for $\bar{F}$ in the same way that cells containing **1** are looped to give an expression for F. Use this fact to show that the truth table given in table 4.3 is described by the following boolean relationship.

$$F = \overline{\bar{A} \cdot C + \bar{A} \cdot \bar{B} \cdot D + \bar{B} \cdot C \cdot D}$$

Bibliography

1 M. Karnaugh. The Map Method for Synthesis of Combinational Logic Circuits. *Trans. Am. Inst. elect. Engrs*, 72 (1953), 593-9.
2 E. W. Veitch. A Chart Method for Simplifying Truth Functions. *Proc. Ass. Comput. Mach.* (1952), 127-33.
3 E. Mendelson. *Boolean Algebra and Switching Circuits*, Schaum Outline Series, McGraw-Hill, New York (1970), chapter 4.
4 J. H. Hill and G. R. Peterson. *Introduction to Switching Theory and Logical Design*, Wiley (1968), chapter 6.
5 V. W. Quine. The Problem of Simplifying Truth Functions. *Am. math. Mon.*, 59 (Oct. 1952), 521-31.
6 E. J. McCluskey. 'Minimisation of Boolean Functions'. *Bell Syst. tech. J.*, 35 (Nov. 1956) 1417-44.

5

Bistable Systems

Logic systems discussed up to this point have had the property that the logical state of the output at a particular point in time was completely determined by the logic levels at the inputs at that time. We now turn our attention to systems in which the logical state of the output at a particular time may be determined by the logic level which existed at the input at some previous time. Putting this another way, we might say that systems of this type are primitive memory elements since observation of the output can be used to deduce the state of the input at some time previously. Systems employing these elements are referred to as sequential logic systems.

5.1 The R-S flip-flop

This is one of the more simple systems and a version consisting of cross-coupled NOR gates is shown in figure 5.1 together with the symbol for the circuit. The 'set' and 'reset' points S and R are normally regarded as inputs to the system with Q as the output.

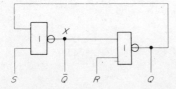 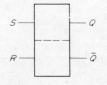

Figure 5.1

To understand the operation of the system, first consider the situation with $R = S = 0$. In this case, it is easily seen that Q can be either **0** or **1** without violating the relationships which must exist between logic levels at the inputs and output of the NOR gates. Figure 5.2 shows the two possible states of the system. Nothing we have said so far can give us any help in determining which of the two possible states is adopted by the system.

BISTABLE SYSTEMS

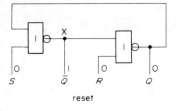 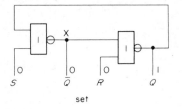

       reset   set

Figure 5.2

If the input state is now changed to $R = 1, S = 0$ this forces the outputs to $Q = 0, \bar{Q} = 1$. No alternative is now possible and the system is said to be reset. Restoring the inputs to $R = 0, S = 0$ will leave the system in the reset condition.

In a similar manner, making the input state $R = 0, S = 1$ forces the outputs to become $Q = 1, \bar{Q} = 0$ and the bistable is then set. Restoring the input condition to $R = 0, S = 0$ leaves the system set with $Q = 1$.

With $R = 0, S = 0$ therefore, observation of the logic level Q will tell us which of R and S was most recently a logical 1. The system 'remembers' the most recent event at this input. The situation $R = 1, S = 1$ has not been considered since this input is not normally regarded as appropriate for this type of system. This is because $R = 1, S = 1$ establishes the condition $Q = 0$ with the logic level at point X normally $\bar{Q}$) also at 0. Restoring the $R = 0, S = 0$ input can then cause either the set or reset condition to be established with no indication as to which state is preferred.

A simple R-S bistable system can also be formed by two NAND gates cross-coupled as shown in figure 5.3. In this case, the input condition $R = 1, S = 1$, allows two possible states and the set or reset state is forced by making the appropriate input S or R a logical 0. In this case, the input $S = 0, R = 0$ gives indeterminate system operation.

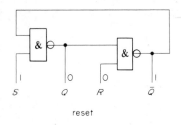 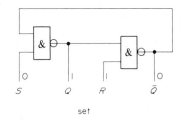

       reset   set

Figure 5.3

Bistable circuit operation is conveniently displayed by means of a truth table. To understand table 5.1 imagine the R, S inputs starting at logical 0 at time t and finishing at logical 0 at time $t + 1$. Between t and $t + 1$, R and S have values indicated by the appropriate columns in the table. The values of Q at t and $t + 1$ describe the circuit operation.

52 INTRODUCTION TO DIGITAL LOGIC

Table 5.1

S	R	Q_t	Q_{t+1}
0	0	0	0
0	0	1	1
0	1	0	0
0	1	1	0
1	0	0	1
1	0	1	1
1	1	0	1
1	1	1	1

5.2 Discrete component circuits

Although this discussion of bistable systems refers essentially to systems either made up from integrated circuit NAND-NOR gates or fabricated completely in integrated circuit form, almost all of what is said will apply to discrete component circuits. As an example of this, figure 5.4 shows a single R.T.L. discrete component NOR gate and indicates how two such gates, cross-coupled, form a discrete component R-S bistable circuit.

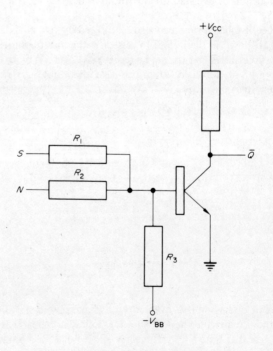

Figure 5.4A

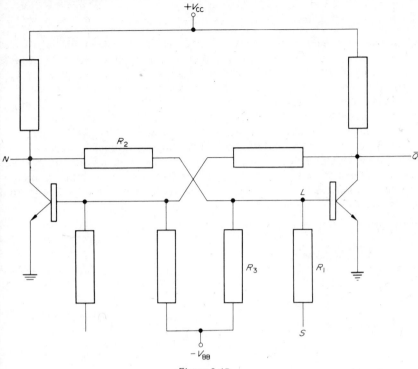

Figure 5.4B

5.3 The clocked R-S flip-flop

This system is a modification of the simple R-S flip-flop which incorporates a third input called the clock input. It is possible to set or reset the system via the S-R terminals only if the clock input is 1. A system and a system symbol is shown in figure 5.5.

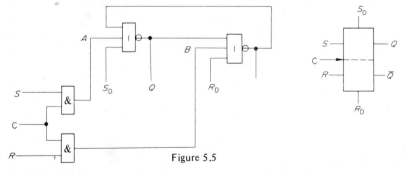

Figure 5.5

It will be seen that A and B will always be 0 so long as C is 0. When C becomes 1, A and B are equal to S and R respectively, allowing the system to be set or reset. Additional inputs S_D and R_D known as direct set and direct reset (or sometimes preset and preclear) may be incorporated to give a set and reset facility which overrides the clock, that is, allows the circuit to be set or reset independently of the clock input.

5.4 Counting elements

When constructing counting systems it is often required to have available bistable elements which change state on the arrival of each clock pulse. Figure 5.6 shows an attempt to achieve this using a clocked R-S flip-flop. If the initial state is $Q = 1$, $\bar{Q} = 0$, this establishes $S = 0$, $R = 1$ at the input so that on the arrival of a clock pulse the circuit is reset resulting in the state $Q = 0$, $\bar{Q} = 1$. The next clock pulse similarly causes the system to be set giving the state $Q = 1$, $\bar{Q} = 0$ once more.

The simple circuit shown is satisfactory only if the R-S flip-flop is of the master-slave variety since an edge-triggered system will continually oscillate between the set and reset conditions so long as the clock input is at 1. In the case of discrete component circuits, edge-triggered flip-flops are used with the addition of steering diodes, capacitors and resistors giving a time delay to avoid the problem of oscillation.

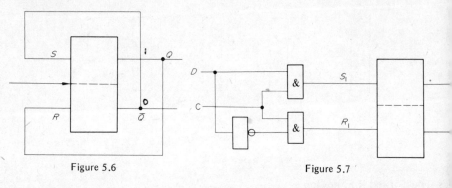

Figure 5.6　　　　　　　　　　Figure 5.7

5.5 The type D flip-flop

This system, often referred to as a 'latch', is shown in figure 5.7. It will be seen that this is a simple modification of the clocked R-S system with only one input in addition to the clock. This circuit avoids problems arising from the possibility of indeterminate operation with the $R = 1$, $S = 1$ input condition. The system shown is the edge-triggered version but the master-slave system is easily implemented in a similar manner to the master-slave R-S flip-flop.

A truth table describing the operation of the system is shown in table 5.2. As with all clocked bistable systems, Q_t refers to the state of the output before the clock pulse and Q_{t+1} refers to the state after the clock pulse. D indicates the value of the input when the clock is at logical 1.

Table 5.2

D	Q_t	Q_{t+1}
0	0	0
0	1	0
1	0	1
1	1	1

BISTABLE SYSTEMS

5.6 The master-slave technique

The clocked R-S flip-flop and type D flip-flop previously described are 'edge-triggered' systems. This means that the set or reset operation is initiated as soon as the clock becomes 1, that is, on the leading edge of a positive-going pulse applied to C. In some situations this can give rise to timing problems and a master-slave system may be a more satisfactory alternative. Figure 5.8 shows the complete system and figure 5.9 shows a simplified version.

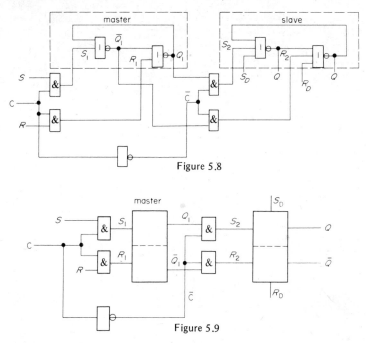

Figure 5.8

Figure 5.9

The symbol for the master-slave clocked flip-flop is identical to that for the edge-triggered version. It will be seen from figure 5.9 that the master flip-flop can only be set or reset when C is 1, that is, after the leading edge of a pulse applied to C. At this time, inputs to the slave flip-flop are disabled (that is, at logical 0) since $\bar{C}$ will be 0. When C returns to 0, that is, after the trailing edge of the pulse at C, the slave flip-flop will be forced to the same state as the master since $\bar{C} = 1$ at this time. The system is therefore 'primed' on the leading edge of the clock pulse when the master flip-flop is set or reset. On the trailing edge of the clock pulse, the logical state of the output of the master is transferred to the output of the slave which is the output of the system.

5.7 The J-K flip-flop

The J-K flip-flop is a system which combines the capabilities of the clocked R-S bistable and the counting element. As with the counting element, since feedback from output to input is involved, the circuit should be implemented using a master-slave technique. Figure 5.10 shows the basic arrangement and figure 5.11 shows the system in more detail, indicating the master and slave elements.

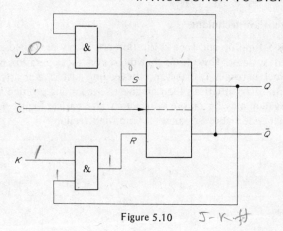

Figure 5.10

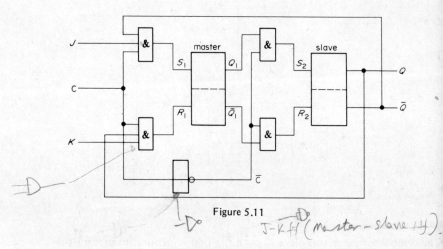

Figure 5.11

The mode of operation is straightforward and it is fairly easy to establish table 5.3.
 Suppose for example, $J = 0$, $K = 1$ and $Q_t = 1$. When the clock input goes to 1, then $S_1 = 0$, $R_1 = 1$ since $Q_t = 1$. The master flip-flop is therefore reset. As soon as the clock input returns to 0, the slave flip-flop is reset by the normal master-slave

	Table 5.3				Table 5.4	
J	K	Q_t	Q_{t+1}	J	K	Q_{t+1}
0	0	0	0	0	0	Q_t
0	0	1	1	0	1	0
0	1	0	0	1	0	1
0	1	1	0	1	1	$\bar{Q}_t$
1	0	0	1			
1	0	1	1			
1	1	0	1			
1	1	1	0			

BISTABLE SYSTEMS

operation. Closer examination indicates that whenever $J = K = 0$, the output state of the flip-flop is the same before and after the clock pulse; that is, $Q_t = Q_{t+1}$. If, on the other hand, $J = K = 1$, the output state after the clock pulse is always the inverse of the output before the clock pulse; that is, the system behaves as a counting element with this input combination. For the two possible remaining input states with $J \neq K$, it will be seen that the system behaves as a clocked R-S flip-flop with the J input having an identical effect as S and the K input corresponding to R. A shortened form of the truth table is shown in table 5.4.

5.8 Integrated circuit bistable elements

A variety of edge-triggered and master-slave flip-flops are now available in integrated circuit form at low cost. The very simple R-S system is most conveniently formed using two integrated circuit NAND-NOR gates, and the versatile master-slave J-K flip-flop can provide most of the facilities required from clocked bistables. The type D latch is the most common of the edge-triggered systems and is available in a variety of configurations.

Problems

1. Show that a J-K flip-flop acts as a counting element if the J input is connected to the $\bar{Q}$ output and the K input is connected to the Q output. Would it be necessary for the J-K flip-flop to be of the master-slave variety for the system to operate satisfactorarily?

2. Figure 5.12 shows the waveform at the clock and D inputs of a type D flip-flop of the type described in section 5.5. Sketch the waveform appearing at the Q output of the flip-flop assuming that Q is 0 initially.

3. Figure 5.13 shows the waveforms at the clock, J and K inputs of a master-slave J-K flip-flop. Sketch the waveform appearing at the Q output of the flip flop assuming that Q is 0 initially.

4. The inputs A, B, C to a logic system change regularly with time in accordance with table 5.5. The output of the system is required to be 0 if $T = n$, $n + 1$ or $n + 7$ ($n = 0, 1, 2, 3$ etc) and 1 otherwise. Show how the system could be implemented using 2 three-input AND gates, 1 two-input OR gate and 1 R-S bistable.

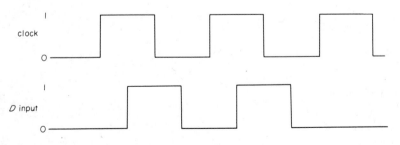

Figure 5.12

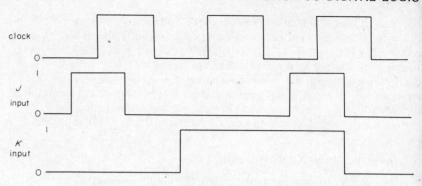

Figure 5.13

Table 5.5

time	inputs		
T	A	B	C
0	0	0	0
1	0	0	1
2	0	1	0
3	0	1	1
4	1	0	0
5	1	0	1
6	1	1	0
7	1	1	1
8	0	0	0
9	0	0	1
etc.			

5 Figure 5.14 shows the waveforms appearing at the preset, preclear, J and K inputs relative to the clock input of a master-slave J-K flip-flop. Sketch the Q output of the flip-flop showing the time relationship with the waveforms given in figure 5.14. Assume that preset and preclear are initiated by a logical **0** signal.

BISTABLE SYSTEMS

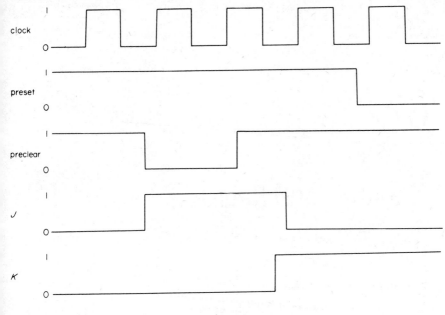

Figure 5.14

Bibliography

1. J. Millman and C. C. Halkias. *Integrated Electronics*, McGraw-Hill, New York (1972), (chapter 17).
2. Texas Instruments. *Designing with TTL Integrated Circuits*, McGraw-Hill, New York (1971) (chapter 7).
3. J. J. Sparkes. *Transistor Switching and Sequential Circuits*, Pergamon, Oxford (1969) (chapter 5).

6

Counters and Registers

6.1 Storage and shift registers

A common requirement in digital computing and instrumentation systems is for some means of storing binary numbers or sequences as logic levels. As will be seen in chapter 9, it is also desirable to be able to manipulate the stored data by shifting to the right or left. Storage and shift registers are collections of bistable elements arranged for the storage and manipulation by left or right shifts of relatively small amounts of binary data. Necessary requirements for such registers are the ability to take in data from a suitable source, preserve the data and have it available for use when required.

Virtually all of the bistable elements described in chapter 5 can be used in storage and shift registers although the clocked circuits are in general more useful. Figure 6.1 shows three type D flip-flops with a common clock line which can be used as a storage register with parallel feed in and out. The data to be stored is presented as logic levels at the D inputs and is stored in the register when a pulse is applied to the clock line. If the bistable elements have preset and preclear facilities, these can also be used to take data into the register.

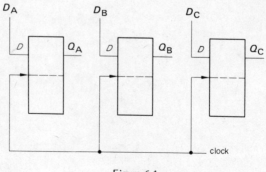

Figure 6.1

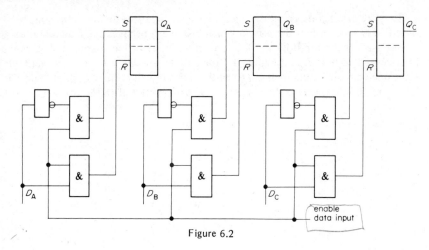

Figure 6.2

Registers which make use of R-S flip-flops require additional gates to facilitate the input of data. Figure 6.2 shows R-S flip-flops used in a three stage parallel input storage register. Data is stored when the 'enable data input' line becomes 1. Arrays of bistable elements suitable for use as storage and shift registers are currently available on single integrated circuit chips. It would not normally be regarded as economical to construct such registers using individual bistable elements.

A simple storage register becomes a shift register if the individual flip-flops are interconnected by allowing the output of one flip–flop to become the input to another. The most straightforward connection consists of a shift-right register and is shown using clocked type D flip-flops in figure 6.3. If the initial state of the register

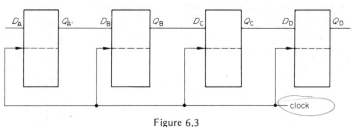

Figure 6.3

is $Q_A = 1$, $Q_B = 0$, $Q_C = 0$, $Q_D = 0$, successive clock pulses will cause the pattern to change in accordance with table 6.1.

Table 6.1

clock	Q_A	Q_B	Q_C	Q_D
0	1	0	0	0
1	0	1	0	0
2	0	0	1	0
3	0	0	0	1
4	0	0	0	0

These remarks concerning the behaviour of the register should perhaps be modified by noting that the way in which bistable A behaves will depend on the circuit used. If the flip-flops are T.T.L. circuits (described in chapter 10), it would be necessary to hold D_A at logical 0 if the pattern in table 6.1 is to be generated since open circuit inputs such as D_A in the diagram will normally adopt logical 1 level. This would generate the pattern shown in table 6.2 if D_A is left open circuit.

Table 6.2

clock	Q_A	Q_B	Q_C	Q_D
0	1	0	0	0
1	1	1	0	0
2	1	1	1	0
3	1	1	1	1
4	1	1	1	1

6.2 Parallel and serial input and output of data

The above consideration leads us naturally to discuss the way in which data can be input to or extracted from a shift register. From the above discussion, it is apparent that data can be input to the register by applying successive input digits to D_A at the same time as successive clock pulses. This is serial data input to the register. It is also apparent that successive digits stored in the register will appear one by one as the logic level at D_D as clock pulses are applied. This provides a means of serial data output from the register.

Parallel data input to a shift register can be achieved by using the preset and preclear facilities of individual flip-flops if they are available or alternatively by using additional gates. Figure 6.4 shows a shift-right register which employs type D flip-flops with preset and preclear capabilities such as the circuits on type 7474 T.T.L. chips. The preset and preclear inputs to most T.T.L flip-flops are activated by logical 0 levels and this is assumed in the circuit of figure 6.4. The preset and preclear inputs also override the D and clock inputs.

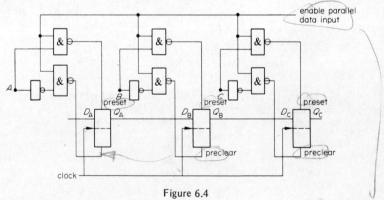

Figure 6.4

The input digits A, B and C are taken into the register when the enable parallel data input line becomes logical 1.

COUNTERS AND REGISTERS

If the flip-flops used in the register do not have preset and preclear facilities, additional gates must be used to enable parallel data input. A suitable scheme in which type D flip-flops are used is shown in figure 6.5.

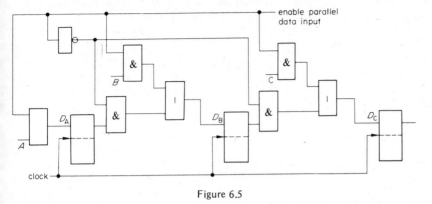

Figure 6.5

Data can only be input to the system shown in figure 6.5 if it is synchronised with a clock pulse. When the 'enable data input' line is taken to logical **1**, the D inputs to the flip-flops becomes equal to the new input data to be stored. If the 'enable data input' line is held at logical **0**, the D input to any flip-flop is equal to the Q output from the preceding stage or in the case of stage 1 to logical **0**.

6.3 Ripple-through counters

Ripple-through or asynchronous counters are sequential logic systems which make use of counting elements of the type discussed in section 5.4. It will be recalled that this type of bistable has the characteristic that the circuit changes state whenever a pulse is received at the input. Using integrated circuit bistables, the usual way of obtaining the desired operation is to use a J-K flip-flop with the J and K inputs held permanently at logical 1. The truth table shown in table 5.4 is then modified to the single relationship $Q_{t+1} = \bar{Q}_t$.

6.4 Frequency division

Suppose we now use a J-K flip-flop connected as a counting element as described above and use a regular train of pulses as the clock input. Assuming that the flip-flop is of the master-slave variety, the logic level at the output of the flip-flop will change on the trailing edge of the clock input. The relationship between the waveforms seen at the input and output of the system is shown in figure 6.6.

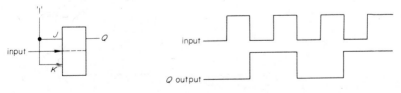

Figure 6.6

It can be seen immediately that the output waveform consists of a train of pulses with a 1 : 1 mark-space ratio and with a repetition rate exactly one half the repetition rate of the input pulse train. This characteristic makes the circuit useful as a frequency dividing element since the frequency of the output waveform is always one half the frequency of the input waveform.

Fairly obviously, the Q output of the bistable could be used as the clock input for a second bistable. The frequency of the waveform at the output of the second bistable will then be equal to the frequency of the system input waveform divided by four. Extending this idea still further, n such bistables connected in cascade will have an output waveform with a frequency equal to the input frequency divided by 2^n.

6.5 Binary counters

In the previous section, the use of bistables connected in cascade as frequency division systems was discussed. We now consider the way in which the state of the individual bistable elements change in such a system with the arrival of each input pulse. Figure 6.7 shows three bistable counting elements connected in cascade. It will be recalled that the Q output of the final bistable will consist of a train of pulses with a frequency equal to the input frequency divided by eight.

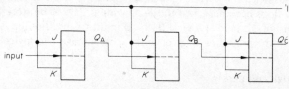

Figure 6.7

Remembering that with master-slave J-K flip-flops connected as shown, the bistables change state on the trailing edge of the input waveform, the waveforms at Q_A, Q_B and Q_C can be deduced. Figure 6.8 shows these waveforms together with the input pulse train. It is assumed that Q_A, Q_B and Q_C are initially at logical 0.

Previously, we have defined the state of a bistable system as set or reset depending on whether the Q output was 1 or 0. This idea can now be extended and we will define the state of a system such as the one shown in figure 6.7 by giving the value of the Q outputs of all bistable elements in the system. Table 6.3 shows the state of the system given in figure 6.7 at the various time intervals defined in figure 6.8 and is obtained directly from figure 6.8.

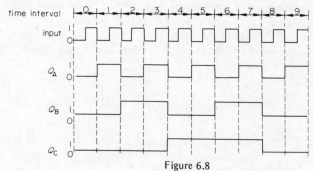

Figure 6.8

COUNTERS AND REGISTERS

Table 6.3

Q_C	Q_B	Q_A	time
0	0	0	0
0	0	1	1
0	1	0	2
0	1	1	3
1	0	0	4
1	0	1	5
1	1	0	6
1	1	1	7
0	0	0	8

Table 6.3 shows that the circuit may adopt any one of eight distinct states. The table also indicates the sequence in which the states occur as the input pulses are received and shows that after eight input pulses, the system returns to the starting state with Q_A, Q_B and Q_C at 0. For this reason the system is sometimes called a scale of eight counter. A further interesting property of table 6.3 emerges if the Boolean variables Q_A, Q_B and Q_C are regarded as binary digits. If these digits are placed in order Q_C Q_B Q_A, this forms three digit binary numbers whose value increases from zero through to seven as the input pulses are received. The binary number Q_C Q_B Q_A therefore indicates the number of pulses received since time 0 (provided this number does not exceed seven). It is for this reason that the system is described as a counter.

Addition of an additional bistable element to the system causes the number of possible states to increase from eight to sixteen and the four bistable output states regarded as binary digits form binary numbers from zero through fifteen. In general, n bistable elements connected in cascade as described above form a scale of 2^n counting system.

Unless the required counting period is very small, it is not usually economical to make up counting circuits from individual bistable elements. Integrated circuit chips are readily available which contain several counting bistable elements connected in cascade to form a binary counter. A four-stage circuit is particularly

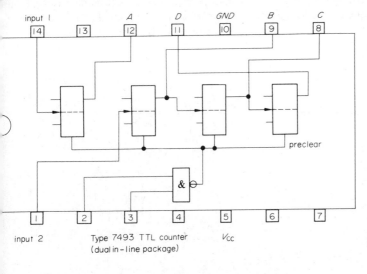

Figure 6.9

useful and figure 6.9 shows details of a four-stage ripple-through counter integrated circuit. The circuit shown is arranged as a three-stage counter with a separate single stage which can be used separately or used with the output feeding the clock input of the three-stage system thus making a four-stage counter. These integrated circuit chips can themselves be connected in cascade to give longer counting periods.

6.6 Modification of counting period

The counting systems described so far are useful for many purposes but are limited by the fact that counting periods are limited to numbers which must be powers of two. Various techniques have been used to design systems which will count with period n where n is any number. Most of these depend on the availability of a direct-reset facility in the bistable elements.

6.7 A scale of six binary counter

The scale of six binary counter is achieved by modifying the scale of eight counter previously described. The technique described here involves the use of a circuit to detect the 'six' state and reset all bistable elements to zero. It will be seen from table 6.3 that after six input pulses, the state of the scale of eight system is normally defined by $Q_C = 1$, $Q_B = 1$, $Q_A = 0$. Rather than adopting this state, we wish the system to adopt the state $Q_C = 0$, $Q_B = 0$, $Q_A = 0$ at this point in the counting cycle.

Examining table 6.3 again, we note that the first time that Q_C and Q_B are simultaneously 1 is after six input pulses. To detect this condition, it is therefore only necessary to connect Q_C and Q_B as the inputs of an AND or NAND gate. Using the very popular ranges of T.T.L logic families, NAND gates are more common than AND gates which means that the output of the gate would become 0 only when the $Q_C = 1$, $Q_B = 1$ condition occurs. This is convenient in many cases since many integrated circuit chips which contain bistable elements have direct-reset inputs which are activated by a logical 0 level. This means that as long as the direct-reset input is at 1, the bistable operates normally. When the direct-reset becomes 0, this overides all other inputs and the bistable is reset. Figure 6.10 shows the circuit for a three-stage binary counter modified as a scale of six circuit. The direct-reset input operates as described above.

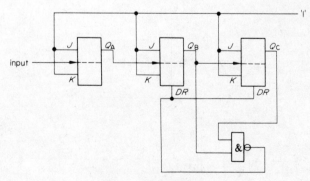

Figure 6.10

COUNTERS AND REGISTERS

The counting sequence is shown in table 6.4.
Figure 6.11 shows the waveforms appearing at Q_A, Q_B and Q_C.

Table 6.4

	Q_C	Q_B	Q_A	time
	0	0	0	0
	0	0	1	1
	0	1	0	2
	0	1	1	3
	1	0	0	4
	1	0	1	5
transitional – – –	1	1	$\bar{0}$	6
state	0	0	0	

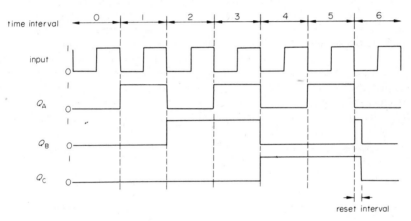

Figure 6.11

The width of the reset interval depends on the time required for the bistable elements to react to the direct-reset input. A figure of 50 ns is typical for the popular T.T.L ranges of integrated circuit bistable elements although circuits which will respond faster are available.

The necessity for the small but finite reset interval places a limitation on the frequency of the input pulse train. Difficulties may also arise if the outputs of the counter are to be decoded to initiate a sequence of events.

6.8 A scale of five counter

This is a further example of the technique described above but this time makes use of an R-S bistable latch to make the direct-reset action more positive. This makes the circuit operation less dependent on delays within the various gates and bistable elements. The counter is shown in figure 6.12.

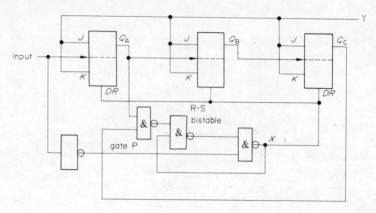

Figure 6.12

Operation of the system is similar to the previous example apart from the effect of the R-S bistable. This bistable is normally in the condition with $X = 1$. When the $Q_A = 1$, $Q_B = 0$, $Q_C = 1$ state is reached at time interval 5, the output of gate P becomes 0 thus resetting the R-S bistable. Note that this occurs on the trailing edge of the pulse at time interval 4 (see figure 6.13). The output of the reset R-S bistable holds the counter bistable element direct-reset lines at zero thus resetting the counter to zero. The R-S bistable remains reset until the clock goes to 1, that is on the next leading edge. At this point, the R-S bistable is reset thus removing the 1 from the direct-reset input to the counter bistable elements. Figure 6.13 shows circuit waveforms.

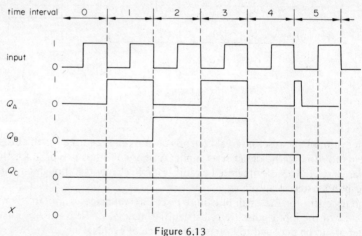

Figure 6.13

6.9 A scale of twelve counter

This additional example of counter design is used to illustrate another variation on the basic design technique. The method used here to modify the basic counting period of a four-stage scale of sixteen counter involves the use of the direct-set or preset facility of the counter bistable elements. During the time period number eleven, on the leading edge of the input waveform all counter bistable elements are set if necessary to make all Q outputs logical **1**. The obvious result of this is that on the trailing edge of the input pulse, that is at the beginning of time interval twelve, all bistables change state and are reset to give the all zero condition. The system is shown in figure 6.14 with appropriate waveforms in figure 6.15.

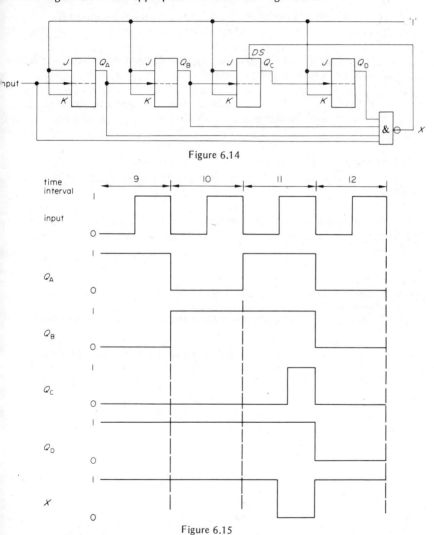

Figure 6.14

Figure 6.15

As may be seen from figure 6.14 and figure 6.15, the preset input required to make $Q_C = 1$ is initiated by using the familiar NAND gate to recognise the state $Q_A = 1$, $Q_B = 1$, $Q_D = 1$ with the system input also providing an input to the NAND gate. The direct-reset input to bistable C therefore remains at **1** until the system input goes to logical **1**, that is on the leading edge of the input pulse.

Problems

1. Design counters with scales of seven and thirteen using the method given in section 6.7.

2. Design counters with scales of nine and eleven using the method given in section 6.8.

3. Design counters with scales of seven and thirteen using the method given in section 6.9.

4. Show the connection between pins of a type 7493 integrated circuit so that the circuit becomes a scale of twelve counter.

5. In an up-down counter, the state of the counter expressed as a binary number will increase or decrease depending on whether the logic level on a control line C is 0 or 1. How could such a counter be implemented as a ripple-through system?

6. Design a ripple-through counting system in which the states of the counter change in accordance with an excess three B.C.D. number sequence (see appendix A).

7. Design a ripple-through counting system in which the states of the counter change in accordance with a 2421 B.C.D. number sequence (see appendix A).

Bibliography

1. F. J. Hill and G. R. Peterson. *Introduction to Switching Theory and Logical Design*, (1968) Wiley (chapter 9).
2. D. C. Steinback. I.C. Frequency Dividers and Counters. *Electronics Wld*, **80**, (December 1968) 32-5.
3. H. W. Gschwind. *Design of Digital Computers*, (1967) Springer, Wien (chapter 6).
4. R. Henner. COS/MOS MSI Counter and Register Design and Applications. *RCA Applic. Note*, **ICAN-6166**.
5. Fairchild. MSI 9308 Dual 4-Bit Latch. *Fairchild Advanced Logic Book*, Fairchild Semiconductor Company, 569-74.

7
Synchronous Counters

Compared with ripple-through counters, synchronous counters have certain advantages which may be important in some cases. A major advantage is that all flip-flops in the counter change state at the same time thus avoiding the delay in propagating the clock pulse from one stage to the next. Some disadvantages of synchronous systems are apparent however in that more components are often required if a large number of stages are involved. It should also be noted that at some state of the count, the outputs of most or all of the flip-flops may change from 1 to 0 or 0 to 1 simulataneously which can give rise to noise spikes on the supply rail.

Traditionally, the study of synchronous counters concentrates on the most common forms, such as systems whose output states change in a simple binary or B.C.D. pattern (B.C.D. numbers are described in appendix A.) These systems are now readily available at low cost on single integrated circuit chips and the engineer is therefore not often called upon to design them. The treatment here will concentrate on general procedures for designing circuits which will count with any desired sequence of output states. Such procedures in any case can also be applied to deal with the more common types of code pattern. It will be assumed throughout this chapter that all flip-flops used are of the master-slave variety thus avoiding many of the problems of race hazards.

7.1 A simple two-stage counter

The required design procedure is best illustrated by a series of examples starting with a simple two-stage counter employing type D flip-flops. The required outputs of the two flip-flops at each stage of the count are shown in table 7.1.

The characteristics of the type D circuits are shown in table 7.2.

Table 7.1

count	Q_B	Q_A
0	0	0
1	1	1
2	1	0
3	0	1

Table 7.2

D_t	Q_{t+1}
0	0
1	1

Table 7.1 indicates that after the arrival of the first clock pulse, that is at count 1, Q_A is required to be 1. Table 7.2 shows that before the clock pulse arrives, that is at count 0, D_A must therefore be 1. Similarly, since Q_B is required to be 1 at count 1 then D_B must be 1 at count 0. Reference to tables 7.1 and 7.2 will establish the required values of D_A and D_B at other stages of the count. This is summarised in tables 7.3 and 7.4.

Table 7.3

count	Q_B	Q_A	D_B
0	0	0	1
1	1	1	1
2	1	0	0
3	0	1	0

Table 7.4

count	Q_B	Q_A	D_A
0	0	0	1
1	1	1	0
2	1	0	1
3	0	1	0

To produce the desired count sequence, it is necessary to use two logic systems, one associated with D_A, the other with D_B to generate the required logic level at each stage of the count as indicated by tables 7.3 and 7.4. The last three columns of tables 7.3 and 7.4 describe the required operation of these systems and are in fact the truth tables for the systems. It follows directly from this that the required operation is described in boolean terms by the following expressions

$$D_B = \bar{Q}_A \cdot \bar{Q}_B + Q_A \cdot Q_B$$
$$D_A = \bar{Q}_A \cdot \bar{Q}_B + \bar{Q}_A \cdot Q_B$$

Simplification is possible by manipulating these expressions directly or alternatively by using a Karnaugh map technique. In this case either method gives the simplest circuit with little effort. For more complicated counters however, map methods have several advantages and this method will therefore be employed in this case. A map is first prepared showing all possible output states together with the number of the count as shown in table 7.1.

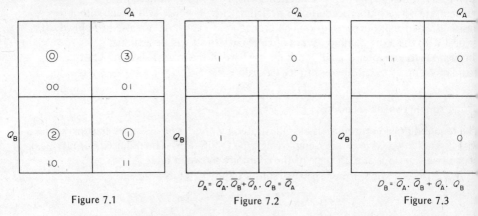

Figure 7.1 Figure 7.2 Figure 7.3

Note that the possible output states are given in the order $Q_B\ Q_A$ in the above diagram, that is the cell in the top right hand corner contains **01** and is labelled 3 since at the count 3, $Q_A = 1$ and $Q_B = 0$. Maps are now prepared to describe the function of the logic circuits associated with D_A and D_B.

SYNCHRONOUS COUNTERS

In figure 7.2, the top right-hand cell contains **0** since this is the cell associated with count 3 and table 7.4 indicates that D_A is required to be **0** at this time.

The system for implementing the operation associated with D_A is easily developed and is shown in figure 7.4. If only one type of logic gate is available, De

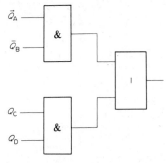

Figure 7.4

Morgan's theorem may be used to eliminate the AND or OR operation as required.

The complete counting system is shown in figure 7.5.

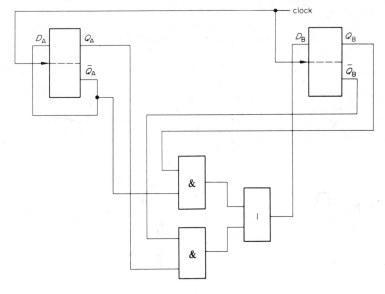

Figure 7.5

7.2 A four-stage counter

The above simple example introduces a design technique which may be developed to deal with most counter design problems. As a more complex design problem, we now consider a four-stage counter whose output states are required to change in accordance with table 7.5. The circuit is to use J-K flip-flops which behave in accordance with the truth table in table 7.6.

Table 7.5

count	Q_D	Q_C	Q_B	Q_A
0	0	0	0	0
1	1	0	0	0
2	1	1	0	0
3	1	1	1	0
4	1	1	1	1
5	1	0	1	0
6	1	0	0	1
7	0	1	1	1
8	0	0	1	1
9	0	0	0	1
10	0	0	0	0

Table 7.6

J_t	K_t	Q_{t+1}
0	0	Q_t
0	1	0
1	0	1
1	1	$\bar{Q}_t$

As a first step in the design procedure, examine the truth table for the bistable elements shown in table 7.6. If a change in the Q output from 0 to 1 is required when a clock pulse arrives then it is necessary for J to be 1 before the clock pulse arrives. The state of K is not important in this case since $K = 1$ or $K = 0$ both result in the required output state $Q = 1$. This means that on the count of 3 for example, $J_A = 1$ and the value of K_A can be 0 or 1, thus ensuring that when the next clock pulse arrives $\dot{Q}_A$ changes from 0 to 1 leaving $Q_A = 1$ on count 4 as required by table 7.5. The values of J and K required to produce the different 'change' or 'no change' situation at the Q output are as follows.

$$Q\ 0 \rightarrow 1\ J = 1$$
$$K = *$$

$$Q\ 1 \rightarrow 0\ J = *$$
$$K = 1$$

$$Q\ 0 \rightarrow 0\ J = 0$$
$$K = *$$

$$Q\ 1 \rightarrow 1\ J = *$$
$$K = 0$$

A table can now be prepared showing the required values for the J and K inputs to the four flip-flops at each stage of the count.

Table 7.7

count	J_A	K_A	J_B	K_B	J_C	K_C	J_D	K_D
0	0	*	0	*	0	*	1	*
1	0	*	0	*	1	*	*	0
2	0	*	1	*	*	0	*	0
3	1	*	*	0	*	0	*	0
4	*	1	*	0	*	1	*	0
5	1	*	*	1	0	*	*	0
6	*	0	1	*	1	*	*	1
7	*	0	*	0	*	1	0	*
8	*	0	*	1	0	*	0	*
9	*	1	0	*	0	*	0	*

SYNCHRONOUS COUNTERS

As in the previous example, a Karnaugh map is now constructed, each cell corresponding to one set of values for Q_A, Q_B, Q_C and Q_D and also indicating the appropriate decimal representation of the count state. Since there are sixteen possible four digit combinations of **0** and **1** and the counter cycles through only ten states, six cells of the map will represent 'don't care' situations.

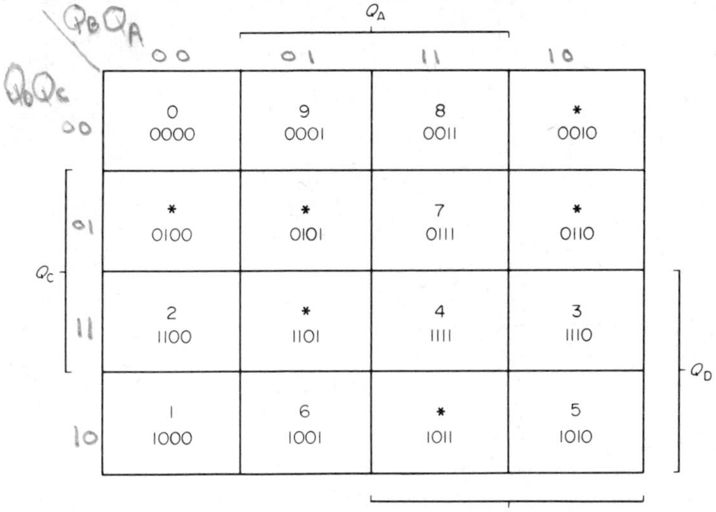

Figure 7.6

The boolean variables in each cell of the map appear in the order Q_D, Q_C, Q_B, Q_A. The third cell from the left on the top row, for example, contains **0011** and is labelled 8 since on the count of 8, $Q_D = 0$, $Q_C = 0$, $Q_B = 1$ and $Q_A = 1$.

The next step in the procedure is to design logic systems to generate the required logic levels for each J and K input at each stage of the count. Since the contents of a cell in figure 7.6 represent one possible set of inputs to these logic circuits, this map may be used as a basis to obtain the minimal design. A map is prepared for each of the four J and four K inputs, indicating the required logic level in the cell labelled with the count number. Looping of the cells now enables the required logical function associated with each map to be obtained in boolean terms. The eight Karnaugh maps with the associated boolean expressions are shown in figure 7.7.

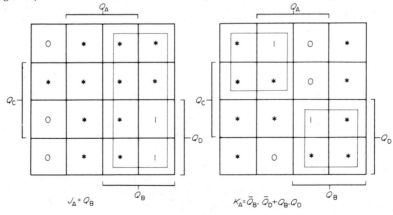

Figure 7.7   [Figure 7.7 continued overleaf

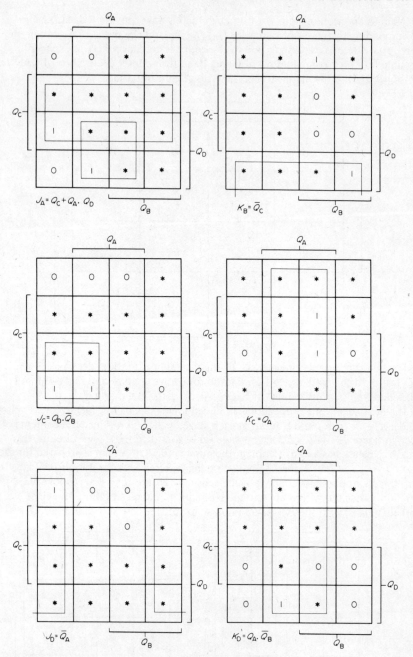

Figure 7.7 continued

Logic systems to generate the various J and K functions using NAND gates only are shown in figures 7.8 to 7.11.

SYNCHRONOUS COUNTERS

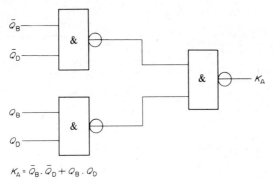

$K_A = \bar{Q}_B \cdot \bar{Q}_D + Q_B \cdot Q_D$

Figure 7.8

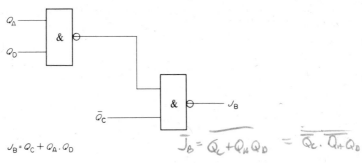

$J_B = Q_C + Q_A \cdot Q_D$

$\overline{\bar{J}_B} = \overline{Q_C + Q_A Q_D} = \overline{\bar{Q}_C \cdot \overline{Q_A Q_D}}$

Figure 7.9

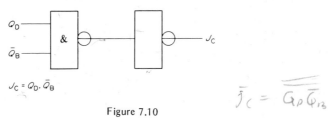

$J_C = Q_D \cdot \bar{Q}_B$

$\bar{J}_C = \overline{Q_D \bar{Q}_B}$

Figure 7.10

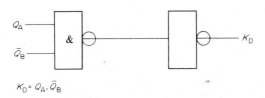

$K_D = Q_A \cdot \bar{Q}_B$

Figure 7.11

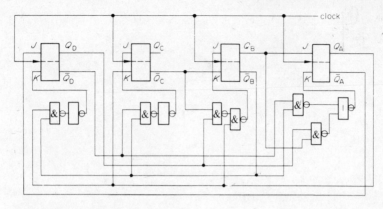

Figure 7.12

7.3 Variable sequence counters

Sometimes counters are required with the facility for modifying the counting sequence by means of an external control signal. Up-down counters are the most common circuits of this type. The sequence of states for a three-stage circuit which generates a Gray code is shown in table 7.8.

Table 7.8

	Q_C	Q_B	Q_A	
	0	0	0	
	0	0	1	
	0	1	1	
count	0	1	0	count
down	1	1	0	up
	1	1	1	
	1	0	1	
	1	0	0	

Suppose the state of the counter is represented by **110**, that is, $Q_A = 0$, $Q_B = 1$, $Q_C = 1$. If the up control is energised, the counter state will become **111** after the next clock pulse. If the down control is energised, the state of the counter will be **010** after the next clock pulse. To implement count patterns of this type, it is necessary to include the control line logic level in the design procedure for the systems associated with the flip-flop input functions.

Consider the design procedure for an up-down counter generating the code pattern shown in table 7.8 using type D flip-flops with characteristics shown in table 7.2. It will be assumed that a control signal C is available which is at **1** when the count up sequence is required and **0** when the count down sequence is required. A table can be prepared indicating the sequence of states for the $C = 1$ and $C = 0$ conditions. Since this table indicates the successive states of the flip-flops, the required values of the D inputs can also be included as shown in table 7.9.

As with the design procedure for simple synchronous counters, the major problem is now to design the logic systems associated with the D input of each flip-flop. The possible inputs to these logic systems must now include C as well as Q_A, Q_B and Q_C, together with the inverses of these quantities. A Karnaugh map is

SYNCHRONOUS COUNTERS

therefore drawn up with a cell now identified not only with the output state of the counter but also identifying the value of C. This map is shown in figure 7.13. The cell in the top right-hand corner of this map for example contains **0010**, the first three digits identifying the states of the flip-flop in the order Q_C, Q_B, Q_A, the final digit indicating the value of C. This cell therefore is associated with the

Table 7.9

Q_C	Q_B	Q_A	C	D_C	D_B	D_A
0	0	0	1	0	0	1
0	0	1	1	0	1	1
0	1	1	1	0	1	0
0	1	0	1	1	1	0
1	1	0	1	1	1	1
1	1	1	1	1	0	1
1	0	1	1	1	0	0
1	0	0	1	0	0	0
0	0	0	0	1	0	0
1	0	0	0	1	0	1
1	0	1	0	1	1	1
1	1	1	0	1	1	0
1	1	0	0	0	1	0
0	1	0	0	0	1	1
0	1	1	0	0	0	1
0	0	1	0	0	0	0

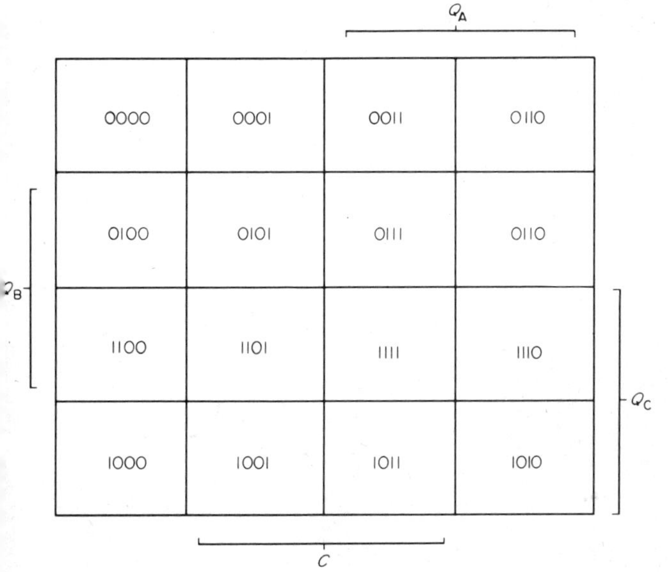

0000	0001	0011	0010
0100	0101	0111	0110
1100	1101	1111	1110
1000	1001	1011	1010

Figure 7.13

condition $Q_C = 0$, $Q_B = 0$, $Q_A = 1$, $C = 0$ implying that the next clock pulse would change the state of the circuit to give $Q_C = 0$, $Q_B = 0$, $Q_A = 0$, since the circuit is 'counting down' in accordance with the lower half of table 7.9 when $C = 0$.

The remaining part of the design procedure follows a similar pattern to that used for counting circuits discussed previously. Karnaugh maps describing the circuits associated with the D inputs to the three stages together with the circuit operation expressed in boolean form are shown in figures 7.14 to 7.16.

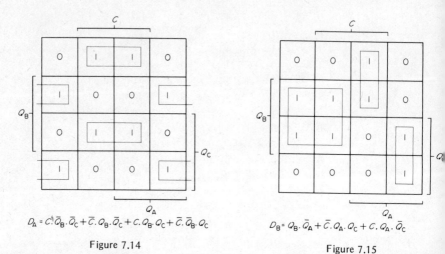

$D_A = C.\bar{Q}_B.\bar{Q}_C + \bar{C}.Q_B.\bar{Q}_C + C.Q_B.Q_C + \bar{C}.\bar{Q}_B.Q_C$

Figure 7.14

$D_B = Q_B.\bar{Q}_A + \bar{C}.Q_A.Q_C + C.Q_A.\bar{Q}_C$

Figure 7.15

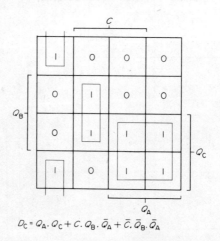

$D_C = Q_A.Q_C + C.Q_B.\bar{Q}_A + \bar{C}.\bar{Q}_B.\bar{Q}_A$

Figure 7.16

SYNCHRONOUS COUNTERS

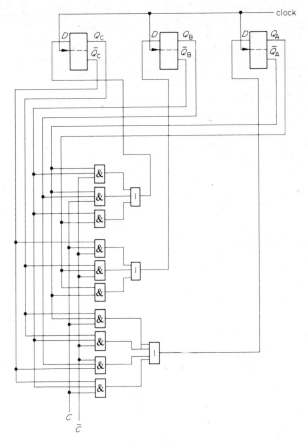

Figure 7.17

Problems

1 Design a synchronous counter which counts with the sequence shown in table 7.10.

Table 7.10

Q_D	Q_C	Q_B	Q_A
0	0	0	0
0	0	0	1
0	0	1	1
0	1	1	1
1	1	1	1
1	1	1	0
1	1	0	0
1	0	0	0
0	0	0	0

etc.

2 Design a synchronous counter in which the counter state changes in accordance with a 2421 B.C.D. number sequence. (see appendix A)

3 Design a synchronous counter in which the counter state changes in accordance with an excess three B.C.D. number sequence. (see appendix A)

4 Table 7.11 shows a *Gray* code counting sequence. Show how the use of exclusive OR gates can greatly simplify the design of a synchronous *Gray* code counter.

Table 7.11

0	0	0	0
0	0	0	1
0	0	1	1
0	0	1	0
0	1	1	0
0	1	1	1
0	1	0	1
0	1	0	0
1	1	0	0
1	1	0	1
1	1	1	1
1	1	1	0
1	0	1	0
1	0	1	1
1	0	0	1
1	0	0	0

Bibliography

1 H. W. Gschwind. *Design of Digital Computers*, Springer, Wien (1967) (chapter 6).
2 J. Millman and C. C. Halkias. *Integrated Electronics*, McGraw-Hill, New York (1972) (section 17.13).
3 R.C.A. Design of Fixed and Programmable Counters Using the RCA CD4018A COS/MOS Presettable Divide-by-N Counter. *RCA Applic. Note*, ICAN-6498.
4 Fairchild. TTL/MSI 9305 Variable Modulo Counter, *Fairchild Advanced Logic Book*, Fairchild Semiconductor Co., 582-96.

8

Simple Sequential Logic Systems

Simple combinational logic systems of the type described in chapter 3 are not adequate if we wish to design logic systems which are required to initiate or respond to a sequence of events or if the response to an input must depend in some way on the inputs which have occurred previous to the current input. In cases such as this, the use of some form of memory element is necessary. Systems described in this section will make use of various types of bistable memory element.

It is possible to discuss systems which make use of memory elements as part of a general mathematical study of sequential systems but we shall here concentrate on a much more simple intuitive approach. The mathematical analysis techniques for sequential systems are useful tools in some cases but the intuitive approach can give satisfactory results even for quite complex systems.

8.1 An electronic combination lock

As a first example we shall consider the design procedure for an electronic combination lock which is to be operated with four push button switches A, B, C and D. These switches must be operated in some given order, say ACBD in order to open the lock.

The first essential in this system is some arrangement which causes the operation of a switch to initiate a change in a logic level. A circuit which makes use of changeover switches is shown in figure 8.1. The outputs of the switch circuits are labelled to correspond with the letter identifying the switch.

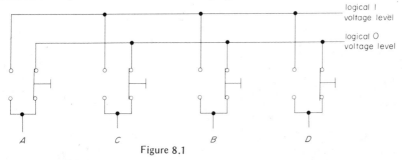

Figure 8.1

83

Now the order in which any two events occur may be determined by means of a single bistable circuit if one event initiates a set signal and the other event a reset signal. In the case of the push button switches A and C, if A resets the bistable and C sets the bistable as shown in figure 8.2 then the state of the bistable can be used to determine which of A and C was pressed most recently. If the bistable is set, that is $Q = 1$, then C was the button most recently pressed and if each switch has been operated just once then the order of operation was A followed by C.

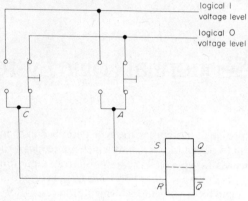

Figure 8.2

The complete design for the lock is now carried through by developing the above ideas a little further. We wish to detect the correct opening sequence which is D follows B, B follows C and C follows A. If each one of these three situations is detected then the lock is required to open. Figure 8.3 shows the complete circuit.

When F becomes 1 this can be used to activate a solenoid to open the lock.

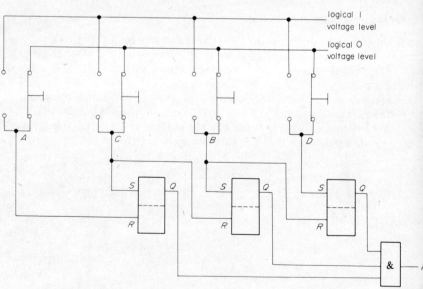

Figure 8.3

SIMPLE SEQUENTIAL LOGIC SYSTEMS

The observant reader will probably detect several deficiencies in the lock system as it has been developed so far. One deficiency in the system is that as designed, button A is redundant, that is, the lock will open if the remaining buttons are pressed in the order CBD whether or not A has been pressed. A second disadvantage in practice would be that the lock could be opened merely by pressing the buttons at random until the correct sequence occurred.

The system shown in figure 8.4 is a more satisfactory system. The four lower R-S bistables will all be set only after each button has been pressed at least once. In this case, therefore, when F and G are both 1 this indicates that all buttons have been pressed at least once and in the correct order which causes H to become 1 thus opening the door. If a situation arises with $G = 1, H = 0$, this indicates that each button has been operated at least once but in the wrong order. This causes H to become 1 thus generating an alarm, the door remaining locked. To lock the door after it is closed, all bistables in the system are reset either by a direct-reset facility or by means of additional gating.

8.2 Lift control systems

Circuits to control the operation of lift systems provide a useful source of examples of circuits involving the use of memory elements. A very simple example will be considered, of a lift operating between two floors.

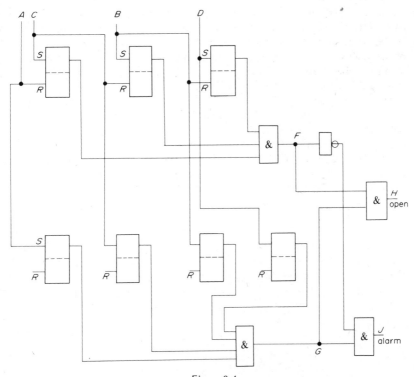

Figure 8.4

We shall suppose that the lift is required to travel between the two floors as demanded by the operation of push button controls. At each floor buttons labelled 1 and 2 are available to control the lift movement. If for example, an operator on floor 1 wishes to transfer goods to floor 2, he first presses button 1 if necessary to bring the lift cage to floor 1. The goods are then loaded, after which the operator presses button 2 causing the lift to travel to floor 2.

As in the previous example, the first essential is to arrange for the operation of the push button switches to generate suitable logic levels to act as inputs to the logic circuit. The push button switches labelled 1 on floors 1 and 2 are connected to logical 0 and logical 1 voltage levels as the switches in the previous example. The switch circuit outputs are used as inputs to an OR gate with the output of the gate identified as $B1$. The two buttons labelled 2 are connected in a similar circuit with the output of the OR gate, this time identified as $B2$.

Pressing a button labelled 1 on either floor causes $B1$ to change from 0 to 1 and pressing a button labelled 2 has a similar effect on $B2$. Some means of indicating that the lift cage has arrived at a particular floor will also be required. Microswitches situated at floors 1 and 2 could be closed by the cage when it reaches a floor. These microswitches can be used in circuits similar to that used with the push button switches to provide input signals to the control logic circuit. We shall denote the outputs from the microswitch circuits by $A1$ and $A2$. $A1$ is 1 when the lift is at floor 1 and, similarly, $A2$ is 1 when the lift is at floor 2. When the lift is travelling between floors, both $A1$ and $A2$ will be 0.

The system to be designed is required to generate an output which connects either the up or the down drive motor to the supply if the inputs $A1$, $A2$, $B1$ and $B2$ require the lift cage to move. In a practical lift system, the situation is usually considerably more complicated than that described here but this simple example provides a useful demonstration of design procedures without becoming too involved with detail.

As a first consideration in the design process, we note that once the lift cage starts to move then it is desirable that it should continue to move until it reaches the desired destination. This means that if the operator presses a control button to direct the lift cage to a particular floor, he would not expect the cage to stop as soon as he removed his thumb from the button. The use of a bistable circuit which is set when the button is pressed is a fairly obvious way of ensuring that the desired conditions causing the lift cage to move persist after the push button switch is released. Since the lift is expected to stop when the desired destination is reached, it is reasonable that the microswitch circuit output at the appropriate floor should be used to reset the bistable.

Let us consider the part of the circuit which generates the control signal for the motor which drives the lift down. It is evident that the signal can only be associated with a demand that the lift should move to floor 1. As soon as floor 1 is reached, the output $A1$ from the microswitch circuit should cause the motor to stop. Figure 8.5 shows a simple approach to the problem.

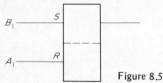

Figure 8.5

SIMPLE SEQUENTIAL LOGIC SYSTEMS

When the button is pressed to direct the lift to floor 1, $B1$ becomes 1 which sets the bistable causing the output to become 1 which connects the down motor to the supply by means of a suitable relay circuit. When the lift reaches floor 1, $A1$ becomes 1 which resets the bistable which causes the output to become 0 thus removing the supply from the drive motor. This appears to be a reasonable solution until one considers what would happen if button 1 was pressed accidentally with the lift already at floor 1 thus producing the situation $A1 = 1$ and $B1 = 1$. The way in which the bistable reacts to this situation will depend on the circuit details of the bistable itself. If it consists simply of two NOR gates connected in the normal manner, a 1 applied simultaneously to the set and reset inputs will result in a 0 output which would be acceptable in this case. More complicated bistable circuits may not behave in this way and the truth table for the bistable should be examined in every case. It should also be noted that with 1 applied to the set and reset terminals, the inverse of the output Q, normally denoted by $\bar{Q}$ may not be available. The simple R-S circuit formed from two NOR gates is a case in point, since with a 1 applied to both the set and reset inputs results in a 0 at the Q and $\bar{Q}$ outputs.

The function performed by a bistable to generate the signal controlling the motor which winds the lift cage downwards must fairly obviously be duplicated by a circuit to control the motor which winds the cage upwards. This will involve a bistable which is set when button 2 on either floor is pressed. A further complication now becomes apparent, however, if we consider the situation in which button 1 is pressed whilst the lift cage is travelling between floors and towards floor 2. With the circuit described so far, both bistables would be set, thus activating the motors winding the lift up and down simultaneously. To guard against this occurrence, it is necessary to ensure that bistable 1 cannot be set if bistable 2 is already set and vice versa. A simple AND or NAND gate is all that is required, as shown in the complete diagram in figure 8.6.

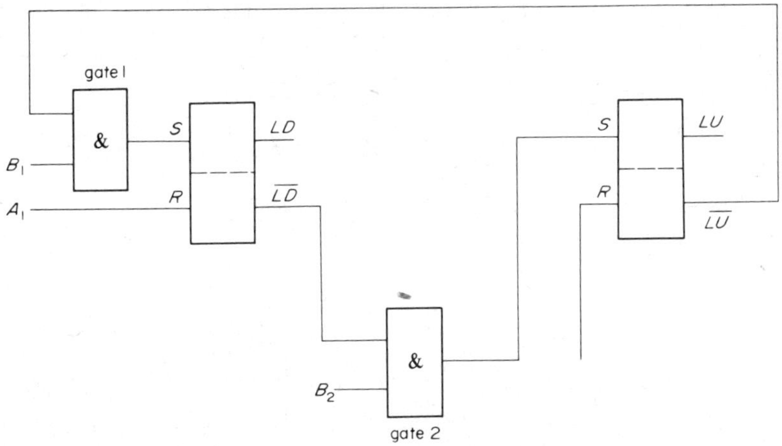

Figure 8.6

The output from gate 1 can only be 1 if $\overline{LU} = 1$, that is if $LU = 0$ implying that the lift down control is not activated. This prevents both bistables being set simultaneously. Gate 2 performs the same function for bistable 2.

8.3 Timing sequences

In many industrial digital systems, an important function of the system is to generate a series of pulses which initiate various control operations at the appropriate times. An important example of this type of system is the control unit of a digital computer. The control unit generates pulses which control the transfer of data between the various functional units within the computer and also enable the sequences of operations involved in arithmetic processes to be performed. The sequences of control pulses required for some of the more simple arithmetic processes are shown in chapter 9.

Control pulse sequences are normally generated by using a clock waveform in conjunction with a counter. To describe the technique a system will be shown to generate the control pulses for the three digit multiplier shown in figure 9.14. The waveforms are shown in figure 9.15 and are reproduced for convenience in figure 8.7. A system which will generate the two pulse trains is shown in figure 8.8.

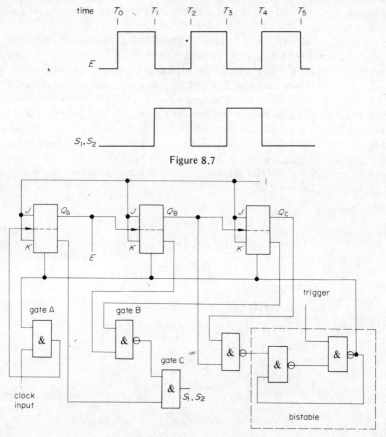

Figure 8.7

Figure 8.8

The E waveform is easily obtained as the output from the first stage of the three stage counter. The sequence of operations is initiated by a negative pulse applied to the trigger input. This changes the state of the R-S bistable and the preclear line of

SIMPLE SEQUENTIAL LOGIC SYSTEMS

the J-K flip-flops is taken to **1** thus allowing the flip-flops to change state if a pulse appears on the clock input to a flip-flop. At the same time, pulses from the clock generator are passed through gate A to the input of the first stage of the counter which therefore commences counting. At the end of the third output pulse at E, the state of the counter changes from $Q_A = 1, Q_B = 0, Q_C = 1$ to $Q_A = 0, Q_B = 1$, $Q_C = 1$. This promptly causes the R-S bistable to change state which takes the preclear line of the J-K flip-flops to **0** thus establishing the conditions $Q_A = 0$, $Q_B = 0, Q_C = 0$ and preventing any further clock pulses from reaching the first stage of the counter. When triggered therefore, the waveform at E consists of three pulses only, which is the required output.

Generation of the $S1, S2$ waveform is a little more complicated. One method would be to observe the states of the counter at the time when $S1, S2$ is required to be 1 and use AND gates to generate 1 at that time. Examination of figure 8.9 shows

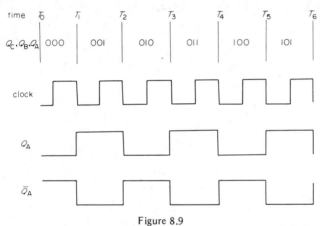

Figure 8.9

that $S1, S2$ is required to be **1** when the state of the counter is $Q_A = 0, Q_B = 1$, $Q_C = 0$ or $Q_A = 0, Q_B = 0, Q_C = 1$. The system shown in figure 8.10 would therefore generate the required output.

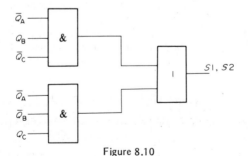

Figure 8.10

As an alternative, we note that the $S1, S2$ waveform is identical to the $\bar{Q}_A$ waveform except in the time interval between T_0 and T_1, that is when $Q_A = 0$, $Q_B = 0, Q_C = 0$. Gates B and C in figure 8.8 ensure that $S1, S2 = \bar{Q}_A$ unless $\bar{Q}_B = 1$ and $\bar{Q}_C = 1$ in which case $S1, S2 = 0$.

As a second design example, consider a system which is required to generate control pulses as shown in figure 8.11 for a control system. Once again, a ripple-through counter may be used in conjunction with an R-S bistable. Figure 8.12 shows the way in which the state of the outputs Q_C, Q_B, and Q_A of the counter stages change.

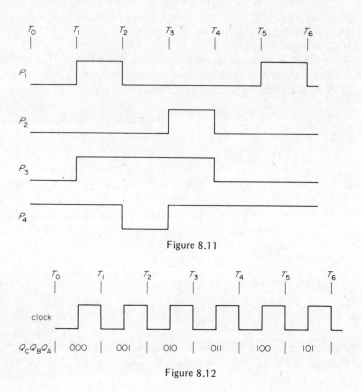

Figure 8.11

Figure 8.12

From figure 8.12, it is apparent that the waveform shown in figure 8.11 will be generated by using a system whose output is 1 between T_1 and T_2 and also between T_5 and T_6, that is when $Q_A = 1$, $Q_B = 0$, $Q_C = 0$ or when $Q_A = 1$, $Q_B = 0$, $Q_C = 1$. This requires a system whose operation is described by the boolean relationship

$$F = Q_A \cdot \bar{Q}_B \cdot \bar{Q}_C + Q_A \cdot \bar{Q}_B \cdot Q_C = Q_A \cdot \bar{Q}_B$$

Systems for generating P_2, P_3 and P_4 are similarly described by the following relationships

$$G = Q_A \cdot Q_B \cdot \bar{Q}_C$$

$$H = Q_A \cdot \bar{Q}_B \cdot \bar{Q}_C + \bar{Q}_A \cdot Q_B \cdot \bar{Q}_C + Q_A \cdot Q_B \cdot \bar{Q}_C = \bar{Q}_C(Q_A + Q_B)$$

$$\begin{aligned}J &= \bar{Q}_A \cdot \bar{Q}_B \cdot \bar{Q}_C + Q_A \cdot \bar{Q}_B \cdot \bar{Q}_C + Q_A \cdot Q_B \cdot \bar{Q}_C + \bar{Q}_A \cdot \bar{Q}_B \cdot Q_C + Q_A \cdot \bar{Q}_B \cdot Q_C \\ &= Q_A + \bar{Q}_B + Q_C\end{aligned}$$

SIMPLE SEQUENTIAL LOGIC SYSTEMS

Any standard method may be used for the simplification of these expressions. Figure 8.13 shows the complete system.

As before, the sequences of pulses are initiated by a negative pulse applied to the trigger input thus changing the state of the R-S bistable. This technique of using a counter together with gates whose output becomes 1 only for certain state(s) of the counter is commonly referred to as 'decoding the counter output'.

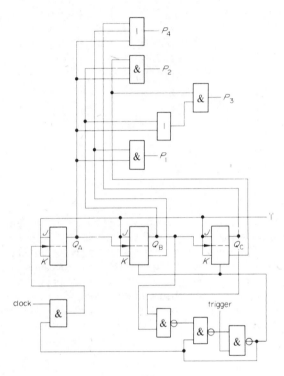

Figure 8.13

8.4 Integrated circuit decoders

Because the decoding operation is required so often in industrial control and instrumentation systems, special purpose integrated circuit chips have been developed for the purpose. Table 8.1 shows the truth table for a type 7442 BCD four line to ten line decoder. It will be seen that an output becomes **0** for only one 8421 B.C.D. character at the input and is **1** otherwise.

Integrated circuits are available for decoding a range of input codes including excess three and Gray code. Three line to eight line or 'one out of eight' decoders are also available as integrated circuits.

Table 8.1

inputs								outputs					
D	C	B	A	0	1	2	3	4	5	6	7	8	9
0	0	0	0	0	1	1	1	1	1	1	1	1	1
0	0	0	1	1	0	1	1	1	1	1	1	1	1
0	0	1	0	1	1	0	1	1	1	1	1	1	1
0	0	1	1	1	1	1	0	1	1	1	1	1	1
0	1	0	0	1	1	1	1	0	1	1	1	1	1
0	1	0	1	1	1	1	1	1	0	1	1	1	1
0	1	1	0	1	1	1	1	1	1	0	1	1	1
0	1	1	1	1	1	1	1	1	1	1	0	1	1
1	0	0	0	1	1	1	1	1	1	1	1	0	1
1	0	0	1	1	1	1	1	1	1	1	1	1	0
1	0	1	0	1	1	1	1	1	1	1	1	1	1
1	0	1	1	1	1	1	1	1	1	1	1	1	1
1	1	0	0	1	1	1	1	1	1	1	1	1	1
1	1	0	1	1	1	1	1	1	1	1	1	1	1
1	1	1	0	1	1	1	1	1	1	1	1	1	1
1	1	1	1	1	1	1	1	1	1	1	1	1	1

8.5 Decoding noise

The two systems described above have depended on the use of ripple-through counters with decoding gates. In some cases, decoding the output of a ripple-through counter can produce an output which is corrupted by a number of noise 'spikes' in addition to the desired pulse(s). This is because the bistable elements in the counter do not change state at the same time, that is, the change in state of the first stage causes the change in the second stage and so on.

Consider a three-stage ripple-through counter with a decoding AND gate whose inputs are $\overline{Q}_A$, Q_B and $\overline{Q}_C$ as shown in figure 8.14. The output of the gate will

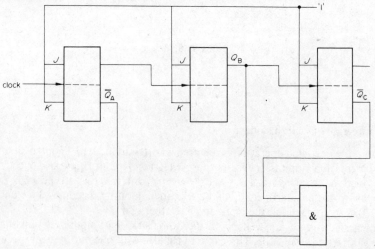

Figure 8.14

SIMPLE SEQUENTIAL LOGIC SYSTEMS

become 1 when the state of the counter is $Q_A = 0$, $Q_B = 1$, $Q_C = 0$. Now we know the counter states are changing in a simple binary sequence. One of the changes will be from $Q_A = 1$, $Q_B = 1$, $Q_C = 0$ to $Q_A = 0$, $Q_B = 0$, $Q_C = 1$, that is from 011 to 100 in the binary counting sequence. Since the changes of state of the three flip-flops will occur one after the other, several 'transition' states will occur before the final state $Q_A = 0$, $Q_B = 0$, $Q_C = 1$ is reached. These transition states are shown in table 8.2.

Table 8.2

Q_A	Q_B	Q_C	
1	1	0	initial state
0	1	0	
0	0	0	transition states
0	0	1	final state

Now one of the transition states is $Q_A = 0$, $Q_B = 1$, $Q_C = 0$ which will be decoded by the AND gate, the output of which will therefore become 1 for the very brief period during which the transition state exists. A 'noise spike' will therefore appear at the output of the decoding gate. For most types of integrated circuits these noise spikes will have very short durations. Because of this, in some cases, the effect will not be noticed. This would be the case for example, if the output of the gate drives a relay circuit which would not respond to such short pulses. On the other hand, the output of the gate may be used as the clock input to a bistable in which case, the noise spike might well cause the bistable to change state at the wrong time.

The problem of decoding spikes is overcome by using a synchronous counter in place of a ripple-through system. Figure 8.15 shows an alternative to the system

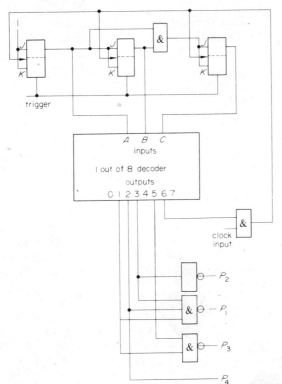

Figure 8.15

shown in figure 8.8 for generating the control pulses shown in figure 8.11. The 1 out of 8 decoder truth table is given in table 8.3. The circuit is triggered by briefly resetting the three flip-flops. This allows clock pulses to reach the flip-flops and the counter commences counting. When the state of the counter becomes $Q_A = 0$, $Q_B = 1$, $Q_C = 1$, no further clock pulses are allowed to reach the flip-flops and the sequence of events is terminated.

Table 8.3

input			outputs							
C	B	A	0	1	2	3	4	5	6	7
0	0	0	0	1	1	1	1	1	1	1
0	0	1	1	0	1	1	1	1	1	1
0	1	0	1	1	0	1	1	1	1	1
0	1	1	1	1	1	0	1	1	1	1
1	0	0	1	1	1	1	0	1	1	1
1	0	1	1	1	1	1	1	0	1	1
1	1	0	1	1	1	1	1	1	0	1
1	1	1	1	1	1	1	1	1	1	0

Problems

1 Design an electronic combination lock operated by three push buttons A, B, C. The lock is required to open if the buttons are operated in the following sequence

> button A operated once
> button B operated three times
> button C operated once

2 Design a lift control system similar to the one described in section 8.2 but operating between three floors. (The system will be considerably more complex than the one described in 8.2).

3 A timing sequence generator is to be designed incorporating a scale of eight counter with outputs Q_A, Q_B and Q_C, Q_A corresponding to the least significant digit of the counting sequence. A logic system decodes the counter output to give an output of 1 for the conditions $Q_A = 0$, $Q_B = 0$, $Q_C = 0$ and the conditions $Q_A = 0$, $Q_B = 0$, $Q_C = 1$. Show the output of the system as the counter cycles through its full counting sequence, indicating where you might expect to see decoding spikes.

4 Even when using synchronous counters, the variation in response of the individual bistable elements might cause very short decoding spikes. Explain why a Gray code counting sequence as shown in table 7.8 might be used to overcome this problem.

SIMPLE SEQUENTIAL LOGIC SYSTEMS

5 A manufacturing process for producing small metal parts performs the following sequence of operations on each part:

   1 trim; 2 smooth; 3 clean; 4 stamp; 5 bend; 6 trim; 7 smooth; 8 clean; 9 polish.

   Design a system to generate control pulses to initiate the various stages in the process.

9

Binary Arithmetic Operations

One of the most important types of digital electronic system is the digital computer. Indeed, it may be argued that much of the pressure for the development of cheaper and faster digital system elements has come from the rapidly expanding computer industry. In this chapter, we consider digital systems for performing arithmetic operations in the binary number system. Such systems of course are important components of digital computers. Readers not familiar with the basic arithmetic operations in the binary number system may wish to refer to appendix A before proceeding with this chapter.

9.1 Binary adders

In logic systems designed to perform arithmetic operations, the logic levels represent binary digits 1 and 0. So far as the design and analysis of such systems is concerned, the techniques described earlier can be used provided one remembers that these techniques represent logic levels as boolean variables rather than binary digits.

One of the simplest digital arithmetic systems is the one bit binary adder. This system adds two single digit binary numbers and is sometimes referred to as a half adder. The system will of necessity have two outputs since the result of adding two single digit numbers may consist of a two-digit number, that is a 'carry' digit may be generated. Table 9.1 shows the possible results of adding two binary numbers A and B and table 9.2 shows the truth table for a system to implement this operation. The two outputs are the sum (S) and the carry (C) digits.

Table 9.1			Table 9.2		
A B		A	B	C	S
0 + 0 = 0 0		0	0	0	0
1 + 0 = 0 1		1	0	0	1
0 + 1 = 0 1		0	1	0	1
1 + 1 = 1 0		1	1	1	0

BINARY ARITHMETIC OPERATIONS

From table 9.2, it follows immediately that the S and C outputs are described in terms of the inputs A and B as follows

$$S = \bar{A} \cdot B + A \cdot \bar{B}$$
$$C = A \cdot B$$

The expression for S describes the well known exclusive OR operation. Implementation of the required operations is of course very straightforward.

If two numbers containing more than one digit are to be added, the system for adding the two least significant digits can be exactly as described above. For the more significant digits, however, the possibility of a carry digit appearing from the previous stage of the calculation must be taken into account. This means that a three-input, two-output, system is required. The inputs consist of the two digits to be added together with the carry digit from the previous stage of addition. The truth table for such a system is given in table 9.3. The three-input, two-output system described by table 9.3 is known as a full adder.

Table 9.3

inputs			outputs	
A	B	C	S	R
0	0	0	0	0
0	0	1	1	0
0	1	0	1	0
0	1	1	0	1
1	0	0	1	0
1	0	1	0	1
1	1	0	0	1
1	1	1	1	1

The operation of the adder is described in boolean terms as follows

$$S = \bar{A} \cdot \bar{B} \cdot C + \bar{A} \cdot B \cdot \bar{C} + A \cdot \bar{B} \cdot \bar{C} + A \cdot B \cdot C$$
$$K = \bar{A} \cdot B \cdot C + A \cdot \bar{B} \cdot C + A \cdot B \cdot \bar{C} + A \cdot B \cdot C \quad (9.1)$$

The expression for K can be simplified in several ways using standard techniques. One simplification gives

$$K = A \cdot C + B \cdot C + A \cdot B \quad (9.2)$$

Figure 9.1 shows a system that implements the above relationships directly.

Figure 9.2 shows an interesting alternative method of implementing the binary addition process. This is in fact the scheme used in one section of the T.T.L. type 7482 two bit binary full adder chip. The mode of operation of this system is less obvious than that of the previous example although it is easily seen that the K output conforms with equation (9.2). If we now consider gate G1, one of the inputs is $\bar{K}$ which is given by the following expression

$$\bar{K} = \overline{A \cdot C + B \cdot C + A \cdot B}$$

Applying De Morgans theorem, we have

$$\bar{K} = \overline{(A \cdot C)} \cdot \overline{(B \cdot C)} \cdot \overline{(A \cdot B)}$$

Applying De Morgans theorem again gives

$$\bar{K} = (\bar{A} + \bar{C}) \cdot (\bar{B} + \bar{C}) \cdot (\bar{A} + \bar{B})$$
$$\bar{K} = (\bar{A} \cdot \bar{B} + \bar{A} \cdot \bar{C} + \bar{B} \cdot \bar{C} + \bar{A} \cdot \bar{C} \cdot \bar{B}) \quad (9.3)$$

From this, the output from gate G1 is given by

$$(\bar{A} \cdot \bar{B} + \bar{A} \cdot \bar{C} + \bar{B} \cdot \bar{C} + \bar{A} \cdot \bar{C} \cdot \bar{B}) \cdot B$$
$$= \bar{A} \cdot B \cdot \bar{C}$$

Similarly, the output from gate G2 is $\bar{A} \cdot \bar{B} \cdot C$ and the output from G3 is $A \cdot \bar{B} \cdot \bar{C}$. The outputs of gates G1, G2 and G3 therefore constitute the first three terms in the expression for S given in equation (9.1). The output from gate G4 obviously gives the fourth term in the expression for S and it immediately follows that the output from the NOR gate G5 is $\bar{S}$.

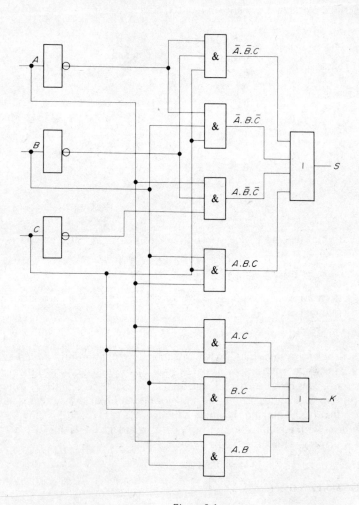

Figure 9.1

BINARY ARITHMETIC OPERATIONS

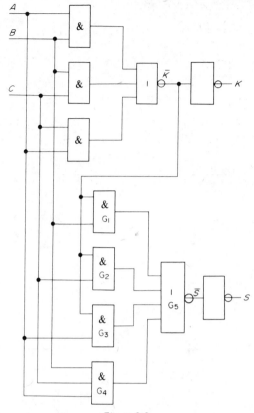

Figure 9.2

9.2 Multidigit parallel adders

We now consider systems for implementing the addition of two binary numbers each of which contains more than one digit. The two numbers to be added will be called A and B. The individual digits of the numbers will be A_0, B_0, A_1, B_1 etc, A_0 and B_0 being the least significant digits.

An obvious approach to the problem of multidigit addition is used in the so called 'ripple carry' system. In practice, the two numbers to be added, that is the addend and augend would be initially contained in storage registers and the result of the addition would be transferred to a storage register on completion. Figure 9.3 shows a system in which the result of the addition is transferred to the register which originally contained A. Systems of the type shown in figure 9.4 give less timing problems if the register stages are master-slave circuits.

Multidigit adders are now widely available as M.S.I. (Medium Scale Integration) circuits. Of the types currently available, the type 4782 two bit adder and type 7483 four bit adder are two of the more common T.T.L. chips. Figure 9.4 shows the system used in the type 7482 two bit adder chip which is based on the adding system shown in figure 9.2.

100 INTRODUCTION TO DIGITAL LOGIC

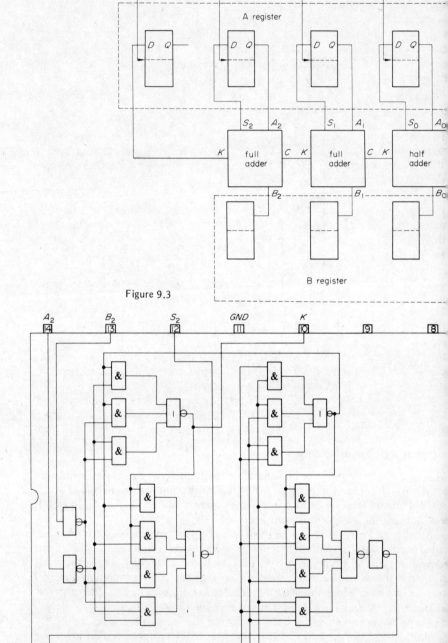

Figure 9.3

Type 7482 TTL. Two bit adder (Dual in-line package)
Figure 9.4

BINARY ARITHMETIC OPERATIONS

The operating speed of ripple-through adders is limited by the fact that before the addition process can be completed at any stage, the correct carry digit must be available from the previous stage. If the delay in any single stage is t, then for an n stage adder, a total time nt must be allowed for the addition process to be completed. In an effort to avoid this time penalty, several techniques have been developed to reduce the ripple time for the carry. One such technique is described below.

9.3 Carry bypass adders

To understand the operation of this system, first consider the carry output of any stage for various inputs. Suppose that the A and B inputs to the stage are both 1. In this case, the carry out, K, will be 1 irrespective of the value of C, the carry in. We can therefore say that this stage generates a carry 1. If on the other hand, both the A and B inputs are 0's the carry out, K, will be 0 even if C, the carry in, is 1. We say that the stage arrests the carry 1 input. The remaining situation to be examined occurs when one input is 0 and the other 1. In this case, if C, the carry input, is 1 then K, the carry out, is also 1. Similarly if C is 0 then K is 0. The stage is said to transfer a carry 1.

The occasions on which every stage in an adder is required to transfer a carry 1 will in general occur very infrequently. In a typical addition process, some stages will generate a carry 1, some will arrest a carry 1 and some will transfer a carry 1. The carry bypass adder takes advantage of this situation by dividing the stages in the adder into groups. A check is made on each group to detect whether a carry 1 input to the group is required to be transferred through every stage in the group. If this is the case, the carry bypasses the group and passes on to the next group. A carry 1 generated within the group will ripple through in the normal manner unless it is arrested. Figure 9.5 shows the general scheme.

The 'carry enable' line must be a 1 only if all stages in the group are required to transfer a carry 1. This can only occur if the two inputs to any stage in the group are unequal, that is, one input is 0 and the other input 1. Table 9.4 shows a truth

Table 9.4

A	B	F
0	0	0
0	1	1
1	0	1
1	1	0

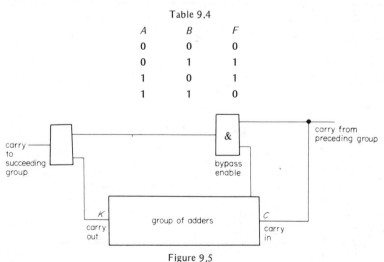

Figure 9.5

table describing this situation for one stage and figure 9.6 shows how the complete bypass enable system could be implemented.

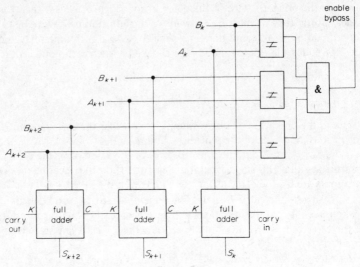

Figure 9.6

9.4 Serial adders

Adding two n digit numbers using the parallel technique described above requires a total of $(n-1)$ full adders and 1 half adder. Because the addition process is carried out at each digit position simultaneously, this method of addition can be comparatively fast. If a considerably longer time can be allowed for the addition to be completed, a serial method of addition which requires one adder and a one-digit store only can result in a considerable saving in cost.

With serial techniques, addition takes place one digit at a time. The inputs to the serial adder are the two digits currently being added together with the carry digit from the previous stage of addition which has been stored. The outputs consist of the sum digit and the carry digit. The carry digit replaces the carry digit from the previous stage of addition in the store.

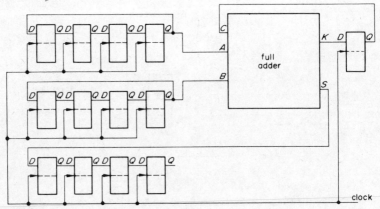

Figure 9.7

Figure 9.7 shows a serial adder system which adds two four-digit numbers. The numbers to be added are initially stored in the two upper shift registers. Clock pulses cause the contents of the registers to be shifted out one digit at a time starting with the least significant digit. At the same time, the digits are recirculated so that after four clock pulses, the contents of the registers are as they were initially. At each stage of the process, the two digits contained in the left-hand stage of the registers are added together with the carry digit from the previous addition which is stored in a clocked bistable store. When the next clock pulse appears, the sum digit is transferred to the sum register and the new carry digit is transferred to the bistable store. The addition process is complete after four clock pulses, the four least significant digits of the sum appearing in the sum register with the least significant digit at the right. The most significant digit of the sum appears in the carry digit bistable store and could be shifted to the sum register if an extra stage were added.

9.5 Signed binary numbers

Whenever arithmetic operations are implemented, sooner or later the problem will arise of how negative numbers are to be represented. Since only two possibilities exist for the sign of a number—plus or minus, the sign can conveniently be regarded as a boolean variable. The complete number is then represented by a set of digits which give the magnitude with one extra digit representing the sign. It is customary to use a sign digit **1** to represent minus and **0** to represent plus.

A variety of methods have been employed to represent the magnitude of a signed number in a logic system. The most obvious technique is to use the direct binary representation of the number irrespective of the sign. Using this notation, plus and minus twenty five would be represented by **011001** and **111001** respectively, the first digit representing the sign in each case. This form of representation is referred to as 'sign and magnitude' notation.

Alternative and more common notations use the 'ones or twos complement' of the magnitude to represent a negative number. Of the two forms of complement notation, twos complement representation is the more common. Ones and twos complements of binary numbers are discussed in appendix A. To represent minus twenty five using twos complement notation, we first note that the sign digit will be 1. Twenty five is represented as a binary number by **11001** and the twos complement of this is **00111**. Minus twenty five would therefore be represented by **100111** using this notation. The ones complement representation differs only in that the ones complement of **11001** is employed, that is **00110**. Minus twenty five in ones complement notation is therefore represented by **100110**.

In a computing system, it is desirable that computations should be possible involving numbers less than and greater than one. This introduces a further complication with regard to the way in which numbers are represented in the system, with the necessity for indicating the position of the binary point.

In modern 'fixed point' computing systems the position of the binary point often is fixed by the programmer immediately to the right of the sign digit which is the first digit in the number representation. Thus **0110** in such a system represents 0.110 or plus three quarters. The positioning of the binary point in this way implies that arithmetic operations only take place between numbers less than one. This apparently rather serious disadvantage is overcome by the use of scale factors associated with numbers involved in computations. The number plus twenty five, for example, which is **011001** in binary form, could be written as **0.11001 x 2^5**. The number represented in the computer is **0.11001** and it is the responsibility of the programmer to keep track of the fact that a scale factor of 2^5 is associated with this number.

Writing programs for arithmetic calculations using a fixed point computing system is inevitably laborious owing to the necessity of keeping track of the scale factors of the numbers involved in the calculation. It is possible to transfer this responsibility to the computer itself which results in a *floating point* system. In a floating point system, the word which represents a number is divided into two parts, a *mantissa* and an *exponent*. The number represented by a word is in the form

$$m \times 2^x$$

where m is the mantissa and x the exponent. The mantissa represents a number with the binary point immediately to the right of the sign digit in the manner described above for a fixed point system. The magnitude of the exponent is of course always greater than one. If we consider a system which uses a word length of eighteen bits, the first twelve of which are used for the mantissa and the remaining six for the exponent then the computer word **011000000000010000** has

and
$$\text{mantissa } m = + \mathbf{0.11000000000}$$
$$\text{exponent } x = + \mathbf{10000}$$

In decimal form therefore, we have

$$m = +0.75$$
$$x = +16$$

and the number is 0.75×2^{16}.

A detailed account of floating point techniques is beyond the scope of this text. In any case, implementation of arithmetic operations with floating point numbers involves the mantissas in much the same sort of operation as with fixed point numbers. The additional operations required which involve the exponents are relatively straightforward.

9.6 Generation of complements

The generation of the ones complement of a binary number involves only the inversion of each digit in the number. Implementation of this operation in either a serial or parallel system is trivial and will not be considered further.

The method for generating the twos complement of a number will vary depending on whether a serial or parallel system is involved. In the case of a parallel system, it will be seen from appendix A that the twos complement of a number is formed by inverting each digit in the number and adding 1 at the least significant digit position. This is easily implemented in a parallel system using an inverter and full adder for each digit position. A system is shown in figure 9.8.

In a serial system it is preferable to start with an alternative method for forming the twos complement. The method is described as follows: examine the digits of the original binary number one at a time starting with the least significant digit. If

BINARY ARITHMETIC OPERATIONS

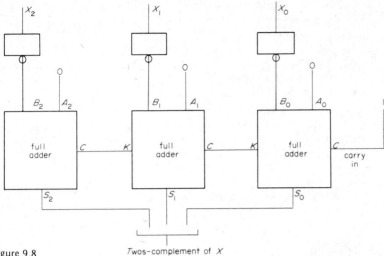

Figure 9.8   Twos-complement of X

the first digit examined is **0**, a **0** appears in the same position in the complement and this continues to be the case with subsequent digits until the first **1** is encountered. When the first **1** is encountered, a **1** is entered in the corresponding position in the complement. For each digit position after this, the digit in the complement is the inverse of the digit in the corresponding position in the original number. Following the method through a few times will soon convince the reader that the correct result is always obtained.

Figure 9.9 shows a serial twos complementer. With this circuit, the master-slave J-K flip-flop is reset at the beginning of the word and remains in this condition provided the input digits are zeros. During this time digits pass through the system unchanged. The first **1** which appears at the input also passes through unchanged but at the end of the clock period leaves the J-K flip-flop set. All subsequent digits input to the system appear inverted at the output. System waveforms are shown in figure 9.10 for an input **0100** giving an output of **1100**. To interpret the waveforms, it must be remembered that the least significant digit is input to the system first and the most significant digit last.

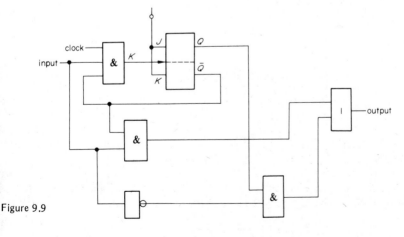

Figure 9.9

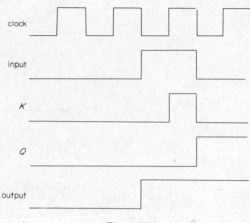

Figure 9.10

9.7 Binary subtraction

Although it is possible to design systems to subtract one binary number from another, very few computers incorporate such systems. This is mainly because the use of complement notation enables negative as well as positive numbers to be added together. A subtraction therefore is easily implemented by complementing the number to be subtracted and adding. Figure 9.11 shows a parallel system for

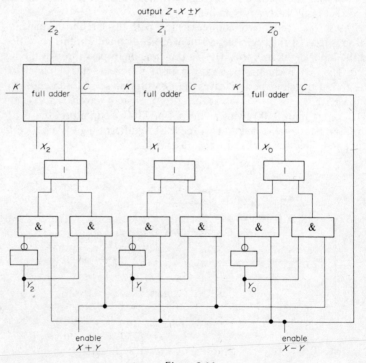

Figure 9.11

BINARY ARITHMETIC OPERATIONS

implementing the $X + Y$ operation using twos complement notation. When $X - Y$ is to be implemented, the twos complement of Y is formed by inverting the digits before they are input to the adders and adding 1 to the least significant position by using a full adder at this position with a carry input of 1. Using twos complement notation, the sign digits should be input to the system along with the other digits. The result will be in correct twos complement form if it is negative and the system can handle positive or negative X and Y inputs. It is important to note that if the result exceeds the capacity of the system, that is becomes greater than unity the system shown will give an incorrect result and *overflow* is said to have occurred. Arithmetic systems which are even more versatile than that shown in figure 9.11 are currently available in the form of single integrated circuits.

9.8 Binary multiplication

The multiplication of two binary numbers can be implemented most simply by following the normal manual multiplication technique. An example is shown in figure 9.12 for $X \times Y$ where $X = 1010$ and $Y = 1001$. The method may be described as follows: examine the digits of X one at a time starting at the least significant digit position. If the digit examined is a 1, form a partial product equal to the current value of Y. If the digit is 0, form a partial product of zero. After each stage in the operation Y is shifted left one position. The result is obtained by adding together all partial products when the digits of X have all been examined.

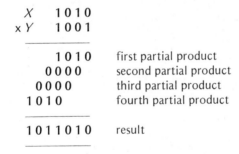

Figure 9.12

In a computer, the machine may be programmed to carry out the individual operations required for multiplication, or alternatively a system may be provided within the machine which performs the required operations in the correct sequence. In the first case, multiplication would be said to be *software* implemented whereas in the latter case, the implementation is by *hardware*. If hardware implementation of the multiplication operation is required, a more economical system results if some small modifications are made to the manual procedure. Rather than store the partial products as they are formed for addition at a later stage, it is usual to add each partial product as it is formed into a register which will finally contain the result of the multiplication. A further modification to the basic method replaces the left shift of X at each stage of the operation by a right shift of the 'results' register. This of course has no effect on the relative positions of the digits in the X register and the digits in the results register when an addition is made into the results register. By avoiding the necessity for left shifts of X, the number of stages in the register holding X is much reduced. Figure 9.13 shows an arrangement for multiplying two three digit numbers giving a six digit result.

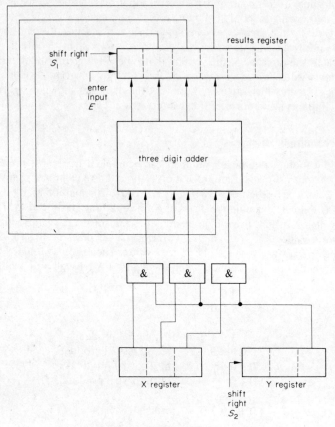

Figure 9.13

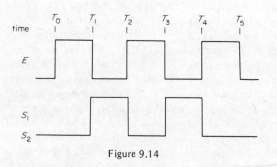

Figure 9.14

Figure 9.14 shows the sequence of timing pulses and table 9.5 shows the register contents at the end of each time interval for $X = 111$, $Y = 101$.

BINARY ARITHMETIC OPERATIONS

Table 9.5
register contents

time	Y	results
T_0	1 0 1	0 0 0 0 0 0
T_1	1 0 1	0 1 1 1 0 0
T_2	0 1 0	0 0 1 1 1 0
T_3	0 1 0	0 0 1 1 1 0
T_4	0 0 1	0 0 0 1 1 1
T_5	0 0 1	1 0 0 0 1 1

It is assumed that the results register is cleared and that X and Y are entered in the appropriate registers before the sequence of operations begins.

Unlike the addition and subtraction systems discussed previously, if the inputs to the system shown in figure 9.14 are negative numbers in twos complement form, the result obtained will not in general be correct. It is fairly easy to check that incorrect results will arise when X and Y are both negative or when X is negative and Y positive. Various techniques exist for multiplication of signed binary numbers using complement notation. A fairly obvious method is to convert the numbers to sign and magnitude form by recomplementing all negative numbers. The multiplication is then carried through and the sign of the result determined separately. If the sign of the result is negative, the multiplier output must then be converted to the correct twos complement form. Another method consists of performing the multiplication with the inputs in twos complement form and modifying the result by adding a correction factor. For a detailed discussion of these and other techniques for multiplication of signed binary numbers, the reader is referred to the more specialised texts listed in the bibliography. The texts listed in the bibliography also give details of techniques for speeding up the multiplication process.

9.9 Binary division

As with multiplication, implementation of the division operation follows the manual or 'paper and pencil' method closely.

```
        X ÷ Y     X = 1 1 0 1 1 1 0 1        Y = 1 0 1 1
   1 0 1 1 ) 1 1 0 1 1 1 0 1 ( 1 0 1 0 0
              1 0 1 1
                1 0 1 1
                1 0 1 1
                0 0 0 0 0 1   remainder
```

Figure 9.15

The procedure demonstrated in figure 9.15 can be described in a manner suitable for implementation by hardware or software as follows: compare Y with a part of

X, starting with the most significant digit of X, then the two most significant digits, then the three most significant digits and so on. At each comparison, if Y is greater than the part of X, a 0 is entered in the result. If Y is less than or equal to the part of X then a 1 is entered in the result and Y is subtracted from the part of X. The procedure is carried through until the total number of comparisons is equal to the number of digits in X. Implementation of the method for the numbers given in figure 9.15 is shown below.

```
compare 1 0 1 1 with 1              larger            enter 0
compare 1 0 1 1 with 1 1            larger            enter 0
compare 1 0 1 1 with 1 1 0          larger            enter 0
compare 1 0 1 1 with 1 1 0 1        smaller           enter 1
       subtract 1 0 1 1 from 1 1 0 1          1 0
compare 1 0 1 1 with 1 0 1          larger            enter 0
compare 1 0 1 1 with 1 0 1 1        equal             enter 1
       subtract 1 0 1 1 from 1 0 1 1            0
compare 1 0 1 1 with 0 0            larger            enter 0
compare 1 0 1 1 with 0 0 1          larger            enter 0
     results 0 0 0 1 0 1 0 0 with remainder 0 1.
```

The major problem in implementing this method directly lies in the complexity of the comparison system required. For the example given above, the comparator would have two four-digit numbers as inputs and would consist of an eight input logic system. Most modern computers operate with numbers containing at least sixteen digits. Comparison of two sixteen-digit numbers would require a thirty-two-input logic system if implemented directly.

To avoid the problems associated with direct comparison, a common technique is to compare the two numbers by subtracting one from the other and checking the sign of the result. After the subtraction if the result is negative it is necessary to restore the number which has been subtracted from to its original value before continuing with the division process. For this reason the method is referred to as 'restoring' division. Description in words of the more complicated arithmetic processes (such as restoring division) becomes somewhat laborious. A more economical method is to use a flow diagram which shows the various operations in a way which enables the sequence to be seen clearly. A flow diagram for the restoring division process is shown in figure 9.16 and a system for implementing the process is shown in figure 9.17. The add/subtract system in figure 9.17 could be similar to the one shown in figure 9.11. It is of course important that in a system such as the one shown in figure 9.17 the various operations should be performed in the correct time sequence. Figure 9.18 shows the sequence of timing pulses when four-digit numbers (that is, three digits plus sign) are being divided.

Table 9.6 lists the register contents at the end of each timing period for X = 0110, Y = 0010. The result is left in the R register at the end of the computation.

As it stands, the system will not give the correct result if the original numbers are negative.

BINARY ARITHMETIC OPERATIONS

Table 9.6
register contents

time	X	Z	R
T_0	0 1 1 0	0 0 0 0	0 0 0 0
T_1	1 1 0 0	0 0 0 0	0 0 0 0
T_2	1 1 0 0	1 1 1 0	0 0 0 0
T_3	1 1 0 0	0 0 0 0	0 0 0 0
T_4	1 0 0 0	0 0 0 1	0 0 0 0
T_5	1 0 0 0	1 1 1 1	0 0 0 0
T_6	1 0 0 0	0 0 0 1	0 0 0 0
T_7	0 0 0 0	0 0 1 1	0 0 0 0
T_8	0 0 0 0	0 0 0 1	0 0 0 1
T_9	0 0 0 0	0 0 0 1	0 0 0 1
T_{10}	0 0 0 0	0 0 1 0	0 0 0 1
T_{11}	0 0 0 0	0 0 0 0	0 0 1 1
T_{12}	0 0 0 0	0 0 0 0	0 0 1 1

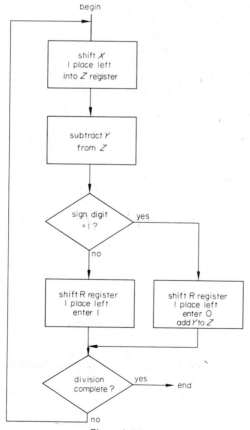

Figure 9.16

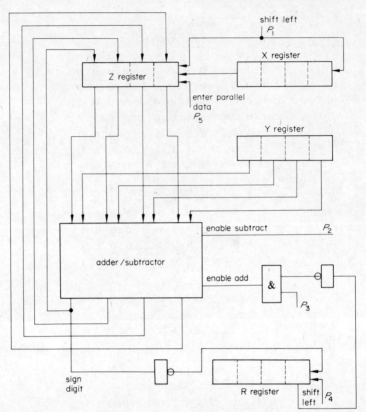

Figure 9.17

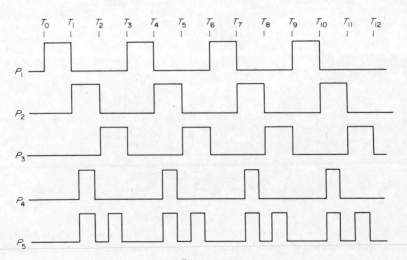

Figure 9.18

BINARY ARITHMETIC OPERATIONS

9.10 The rate multiplier

The rate multiplier is a relatively simple computing circuit which allows a variety of digital functions to be generated at low cost, albeit rather slowly. The mode of operation differs considerably from the arithmetic circuits discussed so far in that the rate multiplier output consists of a train of pulses whose average repetition rate is proportional to the product of the input pulse repetition rate and a second parallel input which represents a binary number. With a few additional components, the circuit can be used to add, subtract, divide, raise to a power, integrate and solve algebraic and differential equations.

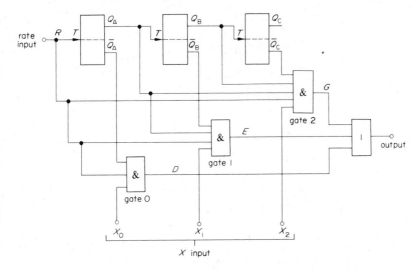

Figure 9.19

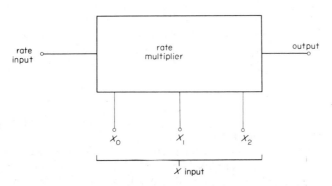

Figure 9.20

Figure 9.19 shows a three-stage rate multiplier circuit with the symbolic representation shown in figure 9.20. The bistable elements in the circuit are type T flip-flops forming a simple ripple-through counter. As has been discussed previously, a J-K flip-flop with $J = K = 1$ will perform the required function. It is not essential for the counter to be asynchronous since the circuit will operate just as well with a synchronous binary counter. The input number X is regarded as a fraction with X_0 as the most significant digit, that is the digit immediately to the right of the binary point.

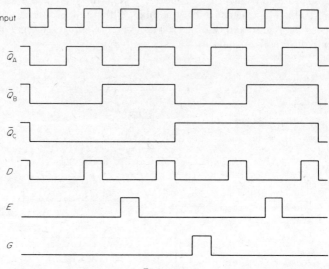

Figure 9.21

The operation of the circuit may be understood by referring to the waveform diagrams shown in figure 9.21. If $X_0 = 1$, the output from gate 0 will be $R \cdot \bar{Q}_A \cdot X_0$, which is the waveform D. Similarly $X_1 = 1$ will result in the output from gate $1 = R \cdot Q_A \cdot \bar{Q}_B \cdot X_1$, which is waveform E and with $X_2 = 1$, the output from gate $2 = R \cdot Q_A \cdot Q_B \cdot \bar{Q}_C \cdot X_2$ will be the waveform G. Now, the repetition rate of the pulse train D is $R_{in}/2$ where R_{in} is the repetition rate of the input. Similarly, the repetition rate of E is $R_{in}/4$ and the repetition rate of G is $R_{in}/8$. If we have, for example, the input $X_0 = 1, X_1 = 0, X_2 = 0$, the output pulse rate is one half the input pulse rate which corresponds with the input regarded as the binary number **0.100** = one half. Since no pulse in D, E and G coincides, if more than one of X_0, X_1 and X_2 are 1, the output pulse rate will be the sum of the two or three pulse rates of D, E and G. Thus if we have the input $X_0 = 1, X_1 = 0, X_2 = 1$, the output pulse rate will be $R_{in}/2 + R_{in}/8 = 5R_{in}/8$ which ties in with the input regarded as a binary number **0.101** = five-eighths.

Instead of regarding the input as a number less than one, we may regard it as an integer. In this case, the output pulse rate would be given by $XR_{in}/2n$ for an n-stage rate multiplier with the input X regarded as an integer; X_0 being the most significant digit as before.

Figure 9.22 shows a system which uses rate multipliers to generate the product of two parallel input numbers. The pulse input to the system is shown in figure 9.23. It is necessary to have the two pulse trains delayed relative to each other to avoid attempting to activate the count up and count down inputs of the counter

BINARY ARITHMETIC OPERATIONS

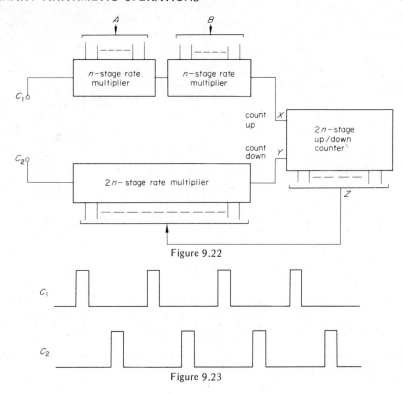

Figure 9.22

Figure 9.23

simultaneously. If we assume that C1 and C2 generate N pulses per unit time, we have for one unit of time

$$X = NAB/2^{2n}$$
$$Y = NZ/2^{2n}$$
$$Z = X - Y = NAB/2^{2n} - NZ/2^{2n}$$
$$Z(1 + N/2^{2n}) = NAB/2^{2n}$$

and if
$$N/2^{2n} \gg 1,$$
$$ZN/2^{2n} \approx NAB/2^{2n}$$

or
$$Z \approx AB$$

Figure 9.24 shows a system for generating the quotient of two parallel input numbers. As before, two waveforms of the type shown in figure 9.23 are required. From figure 9.24 and with N pulses input in unit time we have

$$X = NA/2^n \times 1/2^n$$
$$Y = NB/2^n \times Z/2^n$$
$$Z = X - Y = NA/2^{2n} - NBZ/2^{2n}$$
$$Z(1 + NB/2^{2n}) = NA/2^{2n}$$

and if
$$NB/2^{2n} \gg 1,$$
$$ZNB/2^{2n} \approx NA/2^{2n} \quad \text{or} \quad Z \approx A/B$$

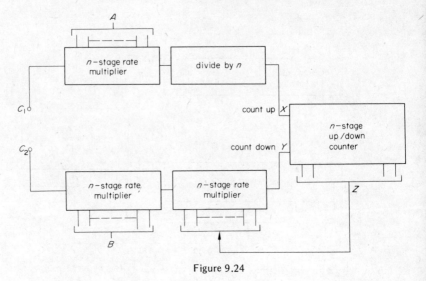

Figure 9.24

The rate multiplier is an interesting newcomer to the field of arithmetic systems, representing as it does a breakaway from the conventional serial and parallel techniques. Although it is doubtful whether rate multipliers will ever replace parallel arithmetic systems in general purpose computers, there are undoubtedly special purpose and dedicated arithmetic situations where the use of rate multipliers will be highly appropriate.

Problems

1 A logic system has two inputs and three outputs. The logic levels at the inputs and outputs represent two and three-digit binary numbers respectively. If n is the number at the input, the number at the output is required to be $2n + 1$. Show a truth table for the system and show how it may be implemented using two gates only.

2 In a system of arithmetic known as modulo three arithmetic, only the symbols **0, 1** and **2** exist and the following relationships hold

$$0 + 0 = 0$$
$$0 + 1 = 1$$
$$1 + 1 = 2$$
$$0 + 2 = 2$$
$$1 + 2 = 0$$
$$2 + 2 = 1$$

Design a logic system in which the modulo three digits are represented by the binary equivalent and which implements the above addition table.

3 Design a serial system to perform addition and subtraction using ones complement notation. Explain how the 'end around carry' operation is implemented in your system.

BINARY ARITHMETIC OPERATIONS

4 Design a non-sequential system which will multiply together two two-digit binary numbers directly.

5 Sketch a block diagram showing a serial multiplier system.

6 Devise a parallel system for generating the square of a binary number.

Bibliography

1 I. Flores. *The Logic of Computer Arithmetic*, Prentice-Hall, Englewood Cliffs, N.J. (1963).
2 R. K. Richards. *Arithmetic Operations in Digital Computers*, Van Nostrand-Reinhold, New York (1955).
3 A. D. Booth. A signed binary multiplication technique, *Q. Jl. Mech. and appl. Math.*, **4**, (1951) 236.
4 O. L. MacSorley. High-speed arithmetic in binary computers, *Proc. Inst. Radio Engrs*, **49**, (Jan. 1961) 67-91.
5 R. G. Saltman. Reducing computing time for synchronous binary division, *Inst. Radio Engrs. Trans. Prof. Group on Electronic Computers*, **V EC-10**, No. 2 (June 1961) 169-74.
6 D. Lewin. *Theory and Design of Digital Computers*, Nelson, London (1972).
7 A. Havasy. Arithmetic arrays using standard COS/MOS building blocks, *RCA Applic. Note* ICAN-6600.
8 R. Phillips. Many digital functions can be generated with a rate multiplier, *Electron. Des.*, **3**, (Feb. 1968) 82-5.
9 Binary rate multiplier, *Texas Instrum. Applic. Rept*, B82.

10

Practical Considerations

10.1 Integrated circuit families

In this chapter an attempt is made to provide information which will assist the logic system designer when translating his design from paper into hardware. At the present time, integrated circuit manufacturers present the designer with a bewilderingly wide array of types of circuit. In addition to designing a system on paper therefore, the engineer may well have to decide which particular range of integrated circuits or 'logic family' is most suitable for the system. This chapter discusses the important features of currently available logic families.

10.2 Packaging of digital integrated circuits

Almost all integrated circuits currently produced are contained in either dual in-line or flatpack packages. Typical dual in-line packages are shown in figures 10.1 and 10.2

Figure 10.1

Figure 10.2

(*Courtesy of Mullard Limited*)

PRACTICAL CONSIDERATIONS

In general, dual in-line packages are easier to handle at the development stage or for manufacturing systems in which the density of components is not critical. Dual in-line packages can be soldered directly on to printed circuit boards or fitted into sockets, which of course simplifies replacement of the circuit. Although it is difficult to generalise, the use of sockets is normally only justified at the development stage since the cost of the socket may well exceed the cost of the integrated circuit chip. The 14 pin dual in-line package shown in figure 10.1 can contain quite complex circuits provided that the number of input and output lines is within the capability of the package. When more input and output pins are required and the circuit is very complex the general shape of the package is retained but is increased in length and/or width as shown.

Flatpacks are most suited to production line techniques and are often welded directly into a printed circuit. Properly executed, this gives an efficient method of production with good reliability. Flatpacks are also the most suitable package for systems in which size is important, that is systems with high package densities. The cost of an integrated circuit in a flatpack package is usually rather greater than the same circuit in dual in-line form.

10.3 Practical system development

When developing a digital system, some form of integrated circuit patchboard is a highly desirable accessory. These patchboards are produced by several manufacturers who advertise regularly in the technical press. They consist of arrays of dual in-line integrated circuit sockets, the pins of which are taken to individual sockets into which leads can be plugged. The integrated circuits are plugged into the dual in-line sockets and interconnections are made between the circuits using patching leads. Construction of this type of patchboard would present few problems to a technician of average ability.

To indicate the logic level at some point within a logic system, a test meter can of course be used in the normal way and is the best instrument if any doubt exists concerning the actual voltage. In many cases, however, an indicator light is more useful. This requires a circuit whose input is connected to the test point and whose output drives a bulb or light emitting diode to give a visual indication if the voltage at the input is in the logical 1 category. Circuits of this type can be built in as part of an integrated circuit patchboard or may be used in a logic probe. A logic probe is usually in the form of a cylinder with a metal probe emerging at one end and a bulb or light emitting diode fitted at the other end. Two power leads also emerge from the body and are clipped to the voltage supply rails of the system being tested. The probe is applied to the test point and the bulb or light emitting diode lights if the voltage is within the logical 1 category. Logic probes are available commercially at low cost.

System layout becomes more important as the speed of the circuits used increases. For very fast logic families such as E.C.L., system layout may be critical, as is discussed in section 10.14. When using high speed logic families such as this, it may not be possible to use a patchboard for development since producing a suitable system layout becomes an integral part of the design procedure. Many of the design techniques are highly specialised, involving multilayer printed circuit boards with interconnections between the layers. Because of these practical difficulties as well as for reasons of cost, very high speed logic circuits should only be used when absolutely necessary.

Even logic circuits of moderate speed such as T.T.L. can give rise to problems associated with layout. Further problems may arise as a result of noise produced on voltage supply lines if the output impedance of the power supply does not remain low at high frequencies. A popular technique is to provide individual printed circuit boards in a system with a smoothed supply at a rather higher voltage than that required by the integrated logic circuits. An integrated circuit voltage stabiliser on each board then brings the supply voltage down to the correct value. The total cost involved in using this technique is little more than providing each board with a voltage supply from a single stabiliser. The problem of transmission of noise on the supply rails between boards is also much reduced.

10.4 Digital circuit parameters

When comparing the suitability of integrated circuit families for a particular application, the parameters defined below should be obtained from the manufacturer's data. Inevitably, the various parameters will assume different degrees of importance depending on the application so that in say a machine tool control system which operates in an electrically noisy environment, noise margin may be the parameter of prime importance whereas in a computing system, propagation delay would probably be of more importance than noise margin.

10.5 Logic levels

Manufacturers will quote voltages within which the high and low voltage logic levels must lie. In particular, figures are quoted for the minimum high output voltage and the maximum low output voltage from any gate in a particular integrated circuit family.

10.6 Threshold levels

These are the input voltages at which the output voltage begins to change from one level to another. Gates will normally have high and low threshold levels. Threshold levels will vary for individual gates in a production run but manufacturers will quote a minimum high threshold level and a maximum low threshold level for an integrated circuit family.

10.7 Noise margin

This parameter is related to the threshold level limits. Noise margin is discussed in chapter 1 as it is related to the transfer characteristics of a gate. The low noise margin is the difference between the maximum low output voltage level and the minimum low input threshold level. The high noise margin is the difference between the minimum high output voltage level and the maximum high input threshold level. If only one figure is quoted for noise margin it will be the smaller of the two.

PRACTICAL CONSIDERATIONS

10.8 Fan out

This is the maximum number of gate inputs to which a gate output may be connected if we are to guarantee that other gate parameters will remain within the specified limits. Fan out depends primarily on the maximum output current which a gate can supply related to the input current requirement.

10.9 Propagation delay time

Two propagation delay times are associated with a gate, these being the rise and fall propagation delay times. The two figures are usually different, and it is often found convenient to express propagation delay as the average of the two figures. If a pulse is applied to the input of an inverting gate, the output will be an inverted pulse with some delay between the input and output pulses. Figure 10.3 shows the relationship between the input and output pulses and defines the rise propagation delay t_{pdr} and fall propagation delay t_{pdf}. The voltage level V_x is midway between the high and low threshold voltage levels.

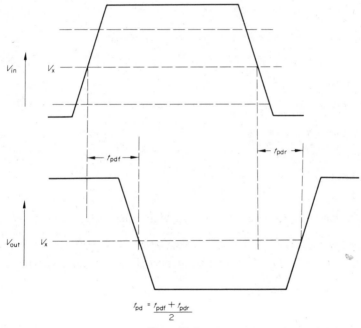

Figure 10.3

It should not be forgotten that in practical systems, the use of gates with very short propagation delay times in a system which does not demand such a rapid response is highly undesirable. If gates with a relatively slow response are used many noise spikes which could be present will be so short that the gate circuit will not be able to respond even though the amplitude of the spikes exceeds the noise margin. Gates with long propagation delay times are also much less critical with regard to the physical layout of a system.

10.10 Wired logic

With some types of circuit, it is possible to perform logic operations merely by connecting the outputs of gates together. Figure 10.4 shows by way of an example, two very simple R.T.L. (resistor–transistor logic) NOR gates with the outputs connected. If either of the two transistors becomes saturated then the output $F = G$ will become low. In boolean terms, the common output of the system is given by $F \cdot G$, since the output is high only if F and G are both high. In terms of the inputs, the output of the system is $(\overline{A + B}) \cdot (\overline{C + D})$.

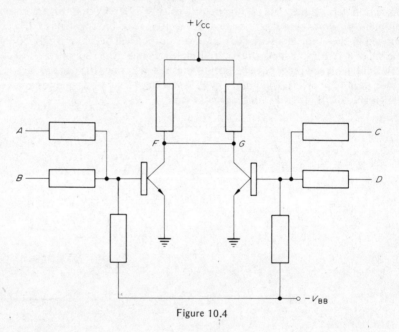

Figure 10.4

Not all logic circuits behave in this way. With some circuits, connecting the outputs of two gates together causes the output to follow the most positive of the individual outputs, thus performing the 'wired OR' operation. Some logic families (in particular T.T.L.) have output circuits which do not permit connection of the outputs of gates thus precluding the use of wired logic. The possibility of wired logic is obviously a desirable feature for an integrated circuit family.

10.11 Resistor-transistor logic

This was the first type of logic circuit to become available in integrated circuit form. Fairly naturally, manufacturers at that time chose to use well-established, discrete component circuits as the basis for their integrated circuits. R.T.L. suffered from a variety of deficiencies, not the least of which was low fan out. From the integrated circuit designer's point of view, R.T.L. has disadvantages since the circuits contain large numbers of resistors compared with transistors. For manufacturing reasons, it is desirable that circuits to be produced in integrated form should have as few passive components as possible. R.T.L. integrated circuits are now obsolete.

PRACTICAL CONSIDERATIONS

10.12 Diode-transistor logic

This could be said to be the first really successful range of integrated circuits and D.T.L. is still in use. Figure 10.5 shows an early type of D.T.L. gate. The circuit comprises a diode AND gate, two level shifting diodes and an output transistor which provides gain and also inverts. The circuit therefore performs the NAND operation.

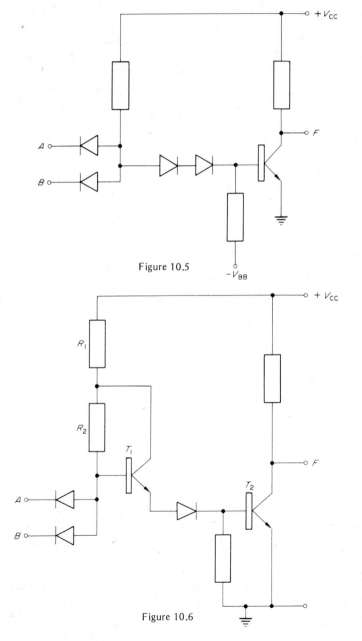

Figure 10.5

Figure 10.6

In the circuit shown in figure 10.6, one of the level shifting diodes has been replaced by the transistor T_1 which reduces the gate input current thus increasing the effective fan out of the gate. This is the configuration of the currently available form of D.T.L.

If all inputs to the circuit shown in figure 10.6 are high, neither of the diodes D_1, D_2 is forward-biased. Current flows through R_1 and R_2 into the base of T_1, driving the transistor into saturation. The saturation of transistor T_1 in turn provides sufficient base drive to hold T_2 in saturation and the output is therefore low. If any one of the inputs is held low, current is diverted from the base of T_1 and flows out of the low input terminal. The removal of T_1 base current removes the base current drive to T_2 which is therefore cut off resulting in the output voltage level becoming high. The circuit therefore performs the NAND operation.

D.T.L. integrated circuits have typical propagation delay times of the order of 25 ns. Fan outs of 8 to 10 and power dissipation of 5 to 10 mW per gate are also typical. An alternative form of this circuit configuration with rather higher resistance values has a reduced power dissipation of typically 1 to 2 mW per gate and a much increased propagation delay time of 50 to 60 ns. Wired logic is possible with D.T.L. integrated circuits.

10.13 Transistor-transistor logic (T.T.L.)

This circuit configuration is often considered as a development of D.T.L. in which the diodes at the input are replaced with the emitter-base junction of a multiple emitter transistor. The multiple emitter transistor, a device rarely seen as a discrete component is easily fabricated on the integrated circuit chip. The T.T.L. gate circuit is shown in figure 10.7.

If all inputs to the circuit shown in figure 10.7 are high, the emitter-base junctions of T_1 will be reverse-biased and current will flow through R_1 and the forward-biased base-collector junction of T_1. This provides sufficient base drive to hold T_2 in conduction which makes the emitter voltage of T_2 sufficiently high to hold T_3 in saturation. Because T_2 is conducting heavily, the collector voltage of T_2 is sufficiently low to hold T_4 cut off. The push-pull output stage of this gate is commonly referred to as a 'totem pole' connection.

If any one of the inputs is taken to a low voltage, current flows out of the corresponding emitter of T_1 since the emitter-base diode is forward-biased. This removes the current drive to the base of T_2 which in turn removes the current drive into the base of T_3. T_3 is therefore cut off. Current now flows through R_2 directly into the base of T_4 thus holding T_4 in saturation and the gate output voltage is therefore high.

Because of the popularity of the basic T.T.L. configuration numerous variations exist to reduce power consumption, increase noise margins or reduce propagation delays compared with the standard configuration. Since improvement of one of these parameters often results in the degradation of another, it is obviously necessary to inspect the complete characteristics of a range of gates before becoming committed to that range. Standard T.T.L. has typical propagation times of 10 ns, power dissipations of 11 mW, and noise margins of 1 V. Some of the variations are listed below.

(1) High speed T.T.L. This range has a higher dissipation than the standard

PRACTICAL CONSIDERATIONS

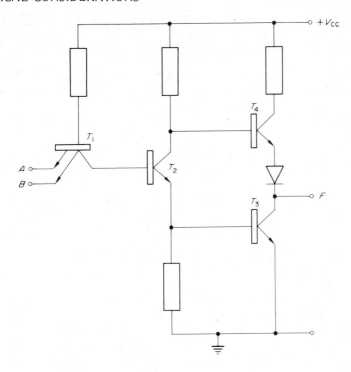

Figure 10.7. T.T.L. Integrated circuit NAND gate.

family with a typical value of 20 mW. The noise margin is similar to that of standard T.T.L. but the propagation delay is reduced to around 6 ns.

(2) Low power T.T.L. The dissipation per gate is reduced to around 1 mW but the propagation delay is increased to a typical value of 35 ns. The noise margin is similar to that of the standard family.

(3) Schottky clamped T.T.L. This is the fastest of the T.T.L. ranges with propagation delays of around 3 ns. Typical power dissipation is 20 mW per gate but the noise margin is worse than standard T.T.L. at approximately 0.3 V.

(4) Tri-state T.T.L. This is a variation of standard T.T.L. in which the output circuit has been changed to allow wired logic which is not possible with the other T.T.L. ranges. This range is also capable of driving lines of up to 3 metres before noise becomes a problem compared with the limit of about 0.6 metres for standard T.T.L.

10.14 Emitter coupled logic

This is a family of logic circuits which differs from families discussed so far in that some transistors in the circuit are operated in a non-saturating mode. Although this mode of operation has some disadvantages, it is undoubtedly the most effective

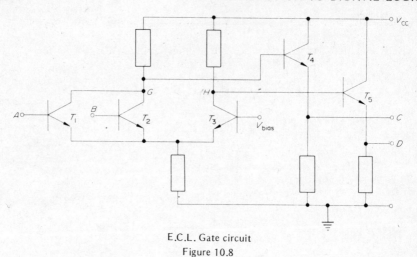

E.C.L. Gate circuit
Figure 10.8

technique when very short propagation delay times are required. E.C.L. finds its main application in very fast computing systems.

If both inputs to the circuit shown in figure 10.8 are low, T_1 and T_2 are both cut off and the emitter voltage is equal the bias input voltage to T_3 plus the emitter base voltage drop. The voltage at point G is therefore greater than the voltage at H since T_3 is conducting whereas T_1 and T_2 are not conducting. The output voltage at C is therefore higher than the voltage at D. T_4 and T_5 act as emitter followers in this circuit.

Suppose either of the two circuit inputs is raised to a higher voltage than the bias input. The emitter voltage will always be determined by the highest of the input voltages and the emitter voltage will therefore rise to become equal to this input voltage plus the emitter-base voltage drop. This reduces the emitter-base voltage of T_3 to a value which is not sufficient to maintain T_3 in conduction. The voltage at H therefore rises and is now higher than the voltage at G. The output voltage at D is therefore now higher than the voltage at C. It will be obvious from this that an OR relationship exists between the gate inputs and output D. The level at C is the logical inverse of the level at D.

The speed of response of this circuit may be judged by the fact that propagation delay times of the order of 2 ns are possible. Even higher speed versions of E.C.L. are available which achieve propagation delay times as low as 1 ns. The power dissipation is relatively high at around 60 mW per gate but the fan out is also quite high with a typical value of 30. A disadvantage of the range is the low noise margin of approximately 0.4 V.

10.15 MOS Integrated circuits

The various types of MOS (metal oxide semiconductor) integrated circuits which make use of MOS field effect transistors represent the most recent arrivals on the digital device scene. Outstanding advantages of MOS circuits are their low power consumption and the fact that they can be fabricated in a very small area on a chip. Because of these advantages, MOS circuits have been widely used for complex systems involving large scale integration (L.S.I.) such as memories for computers.

PRACTICAL CONSIDERATIONS

The level of development is such that it is now possible to produce a complete computer on a single integrated circuit chip using MOS technology.

Both p channel and n channel MOSFETS (metal oxide silicon field effect transistor) are used in the various types of MOS integrated circuits currently available. Figures 10.9 and 10.10 show simple gates which use p channel devices. In

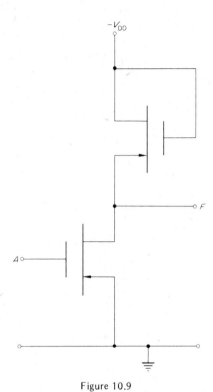

Figure 10.9

figure 10.9 the upper MOSFET acts as a load resistor for the lower MOSFET. If the input is low (in this case negative), MOSFET T_1 is driven into saturation and the output is high (that is, near zero). If the input is high, T_1 is cut off and the output is low. The circuit therefore performs the NOT operation.

In the circuit shown in figure 10.10, if either of the two inputs is low (negative), the associated MOSFET becomes saturated and the output is high (near zero). The circuit therefore performs the positive logic NAND operation.

Because of the very high resistance associated with the gate of a MOSFET, it is possible for the parasitic capacitance between the gate and substrate to act as a temporary store of the voltage level at the gate. This makes 'dynamic' MOS circuits possible. Dynamic operation is used mainly in large circuits such as long shift registers or arrays of bistables used as stores. The important feature of this type of circuit is that a clock is necessary which refreshes the stored charge on the gate-substrate capacitance at regular intervals. Using dynamic operation enables the circuit layout to be made very simple and also reduces the circuit power dissipation.

128   INTRODUCTION TO DIGITAL LOGIC

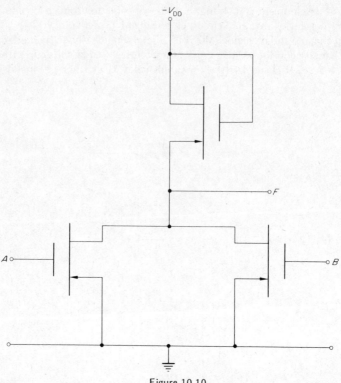

Figure 10.10

10.16 Complementary M.O.S. logic

This family of logic gates is based on the use of n channel and p channel M.O.S. devices employed in a complementary symmetric configuration. Figure 10.11 shows the layout for an inverting or NOT gate. When the input voltage is low, the n channel MOSFET is turned on and the p channel MOSFET is turned off. The resulting output voltage is determined by using the V_D-I_D characteristic of the off p channel MOSFET as a 'load line' superimposed on the V_D-I_D characteristic of the on n channel device; this is shown in figure 10.12. When the input voltage is high, the n channel MOSFET is turned off and the p channel is turned off. Once again, a graphical construction can be used to determine the output; this is shown in figure 10.13.

Figure 10.14 shows a complementary MOS two-input positive logic NOR gate. If A and B are both low, transistors T_3 and T_4 are turned off with transistors T_1 and T_2 turned on. F is therefore high. If A is low and B is high, transistors T_1 and T_3 are on with transistors T_2 and T_4 off. F is therefore low. F is also low if A is high and B is low since transistors T_2 and T_4 are on with T_1 and T_3 off. When A and B are both high, T_1 and T_2 are off with T_3 and T_4 on, which again makes F low. This demonstrates the positive logic NOR operation.

PRACTICAL CONSIDERATIONS 129

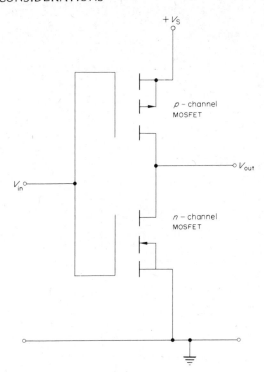

Figure 10.11

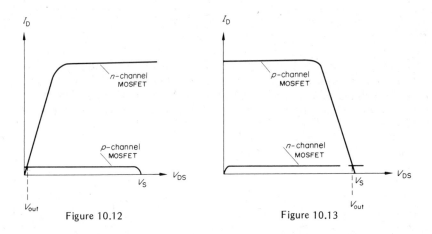

Figure 10.12   Figure 10.13

Examination of figures 10.12 and 10.13 indicates that the high output is only very slightly below the positive supply voltage and the low output voltage very nearly at zero volts. This fact combined with a transfer characteristic having a very sharp 'knee' gives complementary MOS logic circuits an exceptionally high noise margin. A typical transfer characteristic for a gate using a 6 V supply voltage is shown in figure 10.15. The noise margin of complementary MOS gates is typically

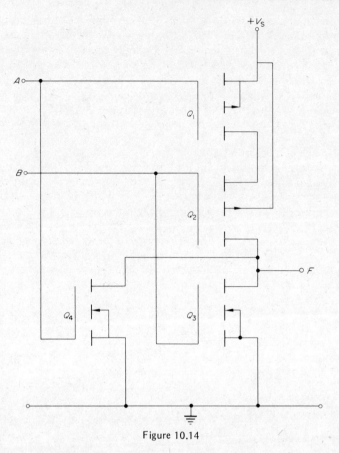

Figure 10.14

40 per cent of the supply voltage. Currently available gates in this family can operate with a wide range of supply voltages from as little as 3 V to 15 V and above. This makes it possible to operate hybrid systems with complementary MOS gates operated from the same power supply as other types of logic in the system.

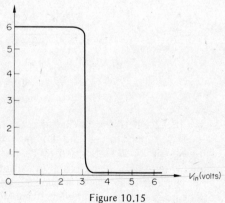

Figure 10.15

PRACTICAL CONSIDERATIONS

A further advantage of this logic family is the extremely low quiescent power consumption. Under quiescent conditions (that is when the input and output are not changing), one or other of the MOSFET's in the circuit shown in figure 10.14 is in the off condition. The extremely high drain to source resistance of MOSFET's results in a low standing current of typically 1 nA at room temperature. The input current of a MOSFET is also extremely low, which means that the fan outs of complementary MOS gates are limited only by the switching speed required. Provided very high speed operation is not required, fan outs of 50 or more are easily attainable.

The input impedance of MOSFET's is mainly capacitive. This means that in spite of the very low quiescent input current, additional current must be supplied to charge the input capacitance when the input voltage is required to change. It is the time required to supply the necessary charge which limits the overall switching time since this charge is normally supplied by the output current of the driving gate. If the number of gates being driven from the output of another gate increases, then the time required to supply the necessary charge to the inputs of these additional gates will also increase. The input capacitance of a MOSFET gate is typically 5 pF at any input. With a load capacitance of 10 pF (equivalent to two inputs), the propagation delay might be around 20 ns rising by 5 ns for each additional 5 pF load in a typical case.

From the above discussion, it will be apparent that the current consumption of a complementary MOS gate will be a maximum when the output is changing from high to low or vice versa. The average power consumption will therefore depend on the total number of such transitions in a given time interval or if the transitions occur at regular intervals on the frequency of these transitions.

Bibliography

1 Texas Instruments. *Designing with T.T.L. Integrated Circuits*, McGraw-Hill, New York (1971).
2 B. Gledhill. The Last Lap—Can You Select the Winner?, *Electron. Engng* (November 1972).
3 R. A. Bishop. Complementary M.O.S. Offers Many Advantages to the Digital Systems Designer, *Electron. Engng*, (November 1972) 67-70.
4 J. Millman and C. C. Halkias. *Integrated Electronics*, McGraw-Hill, New York (1972), chapter 17.
5 R. Sorkin. *Integrated Electronics*, McGraw-Hill, New York (1970).
6 R. H. Crawford. *M.O.S.F.E.T. in Circuit Design*, McGraw-Hill, New York (1967), chapter 5.

Appendixes

Appendix A—systems of numeration

Most people in this country are familiar with several systems of numeration or ways of representing numbers. The most widely used system is decimal numeration based on the arabic numerals 0, 1, 2, 3 etc. but roman symbols such as I, X and C are also widely known and understood.

With the emergence of digital electronic instrumentation, data processing and computation, it has become necessary for engineers to be much more flexible in the approach to numeration and to be competent in handling several different numeration systems.

Number

Number is an abstract concept which is rather difficult to define precisely. Definitions of number given in dictionaries tend to be of limited use for engineering purposes. A concept which many find useful is to consider number as a property which is associated with a set or collection of things and which is independent of the nature of the individual items in the set.

It is important to keep the concept of number separate from the system of numeration. Of two persons asked to write down the number fourteen, one might write 14 while the other might write down XIV. Clearly, 14 and XIV do not represent different numbers although they are different in form. What the original instruction should have said was 'write down *symbol(s)* that *represent(s)* the number fourteen.'

It is particularly important to bear this in mind when working with octal or binary numeration since these systems make use of some of the symbols also used in the decimal system. As an example of this, we may note that five plus seven is always equal to twelve but $5 + 7 = 12$ may or may not be true and is in fact not true when working in some non-decimal systems. This is because twelve is not represented by 12 in all systems of numeration.

APPENDIXES

The decimal system of numeration

This system makes use of ten symbols or digits—0, 1, 2, 3, 4, 5, 6, 7, 8, 9, which are known as hindu-arabic symbols, representing by themselves any whole number between zero and nine. Numbers greater than nine are represented by some combination of two or more digits. In general, a system of numeration consists of a set of symbols together with a rule by which the symbols can be combined together. In the decimal system, the symbols are combined by arranging them in a horizontal line, the contribution which any one digit makes being governed by its position. A decimal point is introduced to enable numbers less than one to be represented also.

For example, 59.36_{10} means

$$5 \times (\text{ten})^1 + 9 \times (\text{ten})^0 + 3 \times (\text{ten})^{-1} + 6 \times (\text{ten})^{-2}$$

Where we understand $(\text{ten})^1$ to mean ten

$(\text{ten})^0$ to mean one

$(\text{ten})^{-1}$ to mean $\dfrac{1}{(\text{ten})}$

$(\text{ten})^{-2}$ to mean $\dfrac{1}{(\text{ten}) \times (\text{ten})}$ and so on.

It is assumed that most people are sufficiently familiar with the basic arithmetic processes of addition, subtraction, multiplication and division in the decimal system.

The octal system of numeration

This system makes use of the hindu-arabic symbols but this time without 8 and 9. Numbers greater than seven must be represented by combinations of two or more digits.

35.73_8 means

$$3 \times (\text{eight})^1 + 5 \times (\text{eight})^0 + 7 \times (\text{eight})^{-1} + 3 \times (\text{eight})^{-2}$$

Where we understand $(\text{eight})^1$ to mean eight

$(\text{eight})^0$ to mean one

$(\text{eight})^{-1}$ to mean $\dfrac{1}{(\text{eight})}$

$(\text{eight})^{-2}$ to mean $\dfrac{1}{(\text{eight}) \times (\text{eight})}$ and so on.

Note that the symbol 8 cannot be included on the above statement since this symbol does not exist in octal numeration. By definition, eight must be represented by 10_8 since this means $1 \times (\text{eight})^1 + 0 \times (\text{eight})^0$ which equals eight.

Eight is known as the *radix* of the octal system just as ten is the radix of the decimal system. Where necessary, the radix can be indicated as a subscript using decimal numeration.

Octal Addition

The technique is similar to that used for decimal addition. Consider the simple addition of 76_8, 22_8 and 15_8.

$$
\begin{aligned}
76_8 &= \text{sixty two} \\
+22_8 &= \text{eighteen} \\
+15_8 &= \text{thirteen} \\
\hline
135_8 &= \text{ninety three}
\end{aligned}
$$

To form the sum, we add five, two and six in the right-hand column which equals thirteen which is 15_8. We write down 5_8 and carry 1_8 to the next column. Adding the 1_8 carried to 7_8, 2_8 and 1_8 gives eleven which is 13_8. We therefore write down 3_8 and carry 1 to the next column as before which gives the result 135_8, which is ninety three.

Octal Subtraction

Once again, the procedure follows that used for decimal subtraction.

$$
\begin{aligned}
32_8 &= \text{twenty six} \\
-17_8 &= \text{fifteen} \\
\hline
13_8 &= \text{eleven}
\end{aligned}
$$

Taking the right-hand column, $2_8 - 7_8$ gives a negative answer. We therefore borrow one eight from the next column (this corresponds to borrowing a ten in decimal arithmetic) and calculate $12_8 - 7_8 = 3_8$. (Remember, $12_8 - 7_8$ means ten − seven). The one borrowed is added to the one in the next column, $3_8 - 2_8$ is calculated giving the final result of 13_8.

Octal Multiplication

The normal procedure of forming partial products is followed. It is important to ensure that the addition of the partial products follows the rules of octal arithmetic.

$$
\begin{aligned}
17_8 &= \text{fifteen} \\
\times\, 21_8 &= \text{seventeen} \\
\hline
17 \quad &\text{first partial product} \\
360 \quad &\text{second partial product} \\
\hline
377 \quad &= \text{two hundred and fifty five}
\end{aligned}
$$

APPENDIXES

Octal Division

This procedure is best illustrated by means of an example

$$375_8 \div 21_8$$

```
    21₈) 375₈ (16₈
         21₈
         ───
         165₈
         146₈
         ───
          17₈  remainder
         ───
```

We first attempt to divide 21_8 into 37_8. Since $21_8 \times 2_8$ is greater than 37, the first figure of the result is obviously 1. $21_8 \times 1$ is now placed below 37_8 and subtracted giving the result 16_8. The next digit is brought down giving 165. We now attempt to divide 165_8 by 21_8 by finding which number when multiplied by 28_8 gives the largest possible number less than or equal to 165_8 (remember this is octal multiplication).

$$\text{Now } 21_8 \times 6_8 = 146_8$$

$$\text{and } 21_8 \times 7_8 = 167_8$$

The next figure in the result is therefore 6. Now 146_8 is placed below 165_8 and subtracted giving 17_8. Since we have no more digits to bring down this is obviously the remainder.

The binary number system

This is a number system which uses the symbols **0** and **1** only. The radix is two and the structure of the system of numeration is similar to that of the octal and decimal systems.

110.11 means

$$\begin{array}{ccccc} 1 & 1 & 0 & 1 & 1 \end{array}$$

$$1 \times (\text{two})^2 + 1 \times (\text{two})^1 + 0 \times (\text{two})^0 + 1 \times (\text{two})^{-1} + 1 \times (\text{two})^{-2}$$

Binary Addition

The basic technique is similar to that used for octal and decimal addition. Consider the following calculation

```
        1 1 0 1
    +     1 0 0
    +     1 1 1
    ───────────
      1 1 0 0 0
```

The sum is formed from the left in the normal manner. Taking the right-hand column, we have one plus zero plus one which equals two which is represented by

10 in this system. We therefore write **0** down and carry one. The next column gives zero plus one plus the one carried which equals two and we once again, write down **0** and carry one. The third column sum is one plus one plus one plus one carried which equals four and is represented by **100**. We therefore write down **0** and carry **10** or two to the next column. The final column sum is one plus the two carried which equals three or **11** which we then write down.

Binary Multiplication and Division

The standard methods of performing multiplication and division apply in the binary system as indicated by the following examples.

```
            1 0 0 1 1              1 0 1 ) 1 0 0 1 1 ( 1 1
        x       1 0 1                      1 0 1
                                          -------
            1 0 0 1 1                      1 0 0 1
            0 0 0 0 0                      1 0 1
                                          -------
          1 0 0 1 1                          1 0 0   remainder
          -----------
          1 0 1 1 1 1 1
```

Binary Subtraction

Although methods used for subtraction in the decimal and octal number system are applicable in the binary system, a subtraction technique using the complement of the number being subtracted has certain advantages. The complement used may be the twos complement, which is the true complement, or the ones complement.

The true complement of a number in any system is formed by subtracting each individual digit from the radix minus one and then adding one to the least significant digit. In the binary number system, this corresponds to changing all **1**'s to **0**'s and vice versa before adding 1 at the least significant digit position. For example, the twos complement of

$$1\ 1\ 0\ 1 = 0\ 0\ 1\ 0 + 1$$
$$= 0\ 0\ 1\ 1$$

To form the ones complement, the step of adding one to the least significant digit is omitted.
That is, the ones complement of **1101** is **0010**.

Radix Conversion

A common requirement is to convert a number in a lower radix (octal or binary) into decimal form. The procedure in this case consists simply of adding the decimal weight of each digit.

Octal to decimal. Convert 173_8 into decimal.

$$173_8 = 1 \times 8^2 + 7 \times 8^1 + 3 \times 8^0$$
$$64_{10} + 56_{10} + 3_{10}$$
$$= 123_{10}$$

APPENDIXES

When converting from a higher to a lower radix, the procedure varies depending on whether the number to be converted is a whole number or a fractional decimal number.

To convert a whole decimal number into one of smaller radix, carry out the following procedure.

(1) Divide the decimal number by the new radix continuously until the result is zero.

(2) Note the remainder at each stage of division and arrange these remainders with the first as the least significant digit and the last as the most significant digit.

For example. Convert 123_{10} into octal.

$$
\begin{array}{ll}
 & \text{remainder} \\
123 \div 8 = 15 & 3 \\
15 \div 8 = 1 & 7 \\
1 \div 8 = 0 & 1 \\
\end{array}
$$

$$\therefore \quad 123_{10} = 173_8$$

As a further example, consider the conversion of 123_{10} into binary form.

$$
\begin{array}{ll}
 & \text{remainder} \\
123 \div 2 = 61 & 1 \\
61 \div 2 = 30 & 1 \\
30 \div 2 = 15 & 0 \\
15 \div 2 = 7 & 1 \\
7 \div 2 = 3 & 1 \\
3 \div 2 = 1 & 1 \\
1 \div 2 = 0 & 1 \\
\end{array}
$$

$$\therefore \quad 123_{10} = 1111011_2$$

To convert a fractional decimal number to one of smaller radix, perform the following operations.

(1) Multiply the decimal number by the new radix continuously (noting the number to the right of the point) until the result is zero.

(2) Arrange the left-hand digits with the first as the most significant digit and the last as the least significant digit.

Note that the process can be stopped when the desired accuracy is achieved, if the product does not go to zero.

For example, convert 0.31_{10} to octal.

$$
\begin{array}{l}
0.31 \times 8 = 2.48 \\
0.48 \times 8 = 3.84 \\
0.84 \times 8 = 6.72 \\
\end{array}
$$

$$\therefore \quad 0.31_{10} = 0.236_8$$

Subtraction Using Twos Complements

To determine $A - B$ where B is greater than A and A, B are binary numbers, the following operations are performed.

(1) Zeros are added to the left of the most significant digit of B until both numbers have the same number of digits.
(2) The twos complement of B is formed.
(3) A is added to the twos complement of B ignoring any carry beyond the most significant digit.

```
For example      1 1 0 1 1₂ − 1 1 1₂ = 27₁₀ − 7₁₀
step (1)                 1 1 1     = 0 0 1 1 1
step (2)    twos complement of 0 0 1 1 1 = 1 1 0 0 0+1
                                         = 1 1 0 0 1
step (3)    1 1 0 1 1
           +1 1 0 0 1
           ─────────
            1 1 0 1 0 0
```

$$\therefore\ 1\ 1\ 0\ 1\ 1_2 - 1\ 1\ 1_2 = 1\ 0\ 1\ 0\ 0_2 = 20_{10}$$

Subtraction Using Ones Complements

To determine $A - B$ where A and B are binary numbers and B is greater than A, the following operations are performed.

(1) Zeros are added to the left of the most significant digit of B to bring the total number of digits equal to the number of digits in A.
(2) The ones complement of B is formed.
(3) A is added to the ones complement of B, the carry beyond the most significant digit being added to the least significant digit of the result.

```
For example      1 1 0 1 1₂ − 1 1 1₂ = 27₁₀ − 7₁₀
step (1)                 1 1 1  − 0 0 1 1 1
step (2)    ones complement of 0 0 1 1 1 = 1 1 0 0 0
step (3)    1 1 0 1 1
           +1 1 0 0 0
           ─────────
            1 1 0 0 1 1

                     1   end around carry
           ─────────
            1 0 1 0 0
```

$$\therefore\ 1\ 1\ 0\ 1\ 1_2 - 1\ 1\ 1_2 = 1\ 0\ 1\ 0\ 0_2 = 20_{10}$$

Binary Coded Decimal Systems

This is a very widely used method of indicating numbers which has a close relationship with the decimal system of numeration as discussed previously. The decimal numeration system uses the ten arabic symbols 0 to 9 to represent numbers

APPENDIXES

between zero and nine. Binary Coded Decimal or B.C.D. codes use sequences of 0's and 1's called code characters to represent these symbols, each code character consisting of four digits. Numbers greater than nine are represented in the same way as in the decimal system but with code characters taking the place of the arabic symbols. Code characters must contain four digits since this is the minimum number of digits required to enable ten unique patterns of 0's and 1's to be formed. The total number of unique patterns of four 1's and 0's is $2^4 = 16$, this means that in any B.C.D. system, some patterns will be unused.

It is interesting at this point to compute the total number of different B.C.D. systems which can be formed. We have sixteen choices for the code character to represent zero, fifteen choices for the character to represent one and so on. The total numbers of B.C.D. systems is therefore

$$16 \times 15 \times 14 \times 13 \times 12 \times 11 \times 10 \times 9 \times 8 \times 7$$

$$= \frac{16}{(16-10)} = 2.9 \times 10^{10}$$

Fortunately, only a very few of this large number of B.C.D. systems are in common use. Four of the more common systems are shown in table A1.

Table A1

decimal number	code character			
	8421	2421	5421	excess three
0	0 0 0 0	0 0 0 0	0 0 0 0	0 0 1 1
1	0 0 0 1	0 0 0 1	0 0 0 1	0 1 0 0
2	0 0 1 0	0 0 1 0	0 0 1 0	0 1 0 1
3	0 0 1 1	0 0 1 1	0 0 1 1	0 1 1 0
4	0 1 0 0	0 1 0 0	0 1 0 0	0 1 1 1
5	0 1 0 1	1 0 1 1	1 0 0 0	1 0 0 0
6	0 1 1 0	1 1 0 0	1 0 0 1	1 0 0 1
7	0 1 1 1	1 1 0 1	1 0 1 0	1 0 1 0
8	1 0 0 0	1 1 1 0	1 0 1 1	1 0 1 1
9	1 0 0 1	1 1 1 1	1 1 0 0	1 1 0 0

To represent 783_{10} in 8421 B.C.D. therefore, we merely use table A1 to find the code character to represent each individual decimal digit.

$$7 - 0\ 1\ 1\ 1$$
$$8 - 1\ 0\ 0\ 0$$
$$3 - 0\ 0\ 1\ 1$$

783_{10} is therefore represented in 8421 B.C.D. by **011110000011**. In a similar manner we could form the excess three B.C.D. representation of the same number as **101010110110**.

Appendix B—logic symbols

Since the time when logic elements first became identifiable entities within the world of electronics, the number and variety of symbols used for their representation has increased steadily with the years. It is unfortunate but perhaps inevitable that no universally accepted system of symbols has come into use. In particular, the Standards Institutions of Europe and the United States have adopted quite widely differing symbols. The question of which logic element symbols are 'best' has, in the past, generated a disproportionate amount of heated argument among engineers and teachers of engineering. It is not the author's intention to add to this discussion. This appendix is included only to summarise the range of symbols used in the text and relate them to other widely used symbols.

One of the first decisions to be made when choosing logic element symbols is the shape used to represent logic gates. At the time of writing, there appears to be a firm intention on the part of British and European Standards Institutions to adopt a rectangular symbol for logic gates (reference B1). For this reason the rectangular shapes of the current British Standard logic element symbols are used in the text.

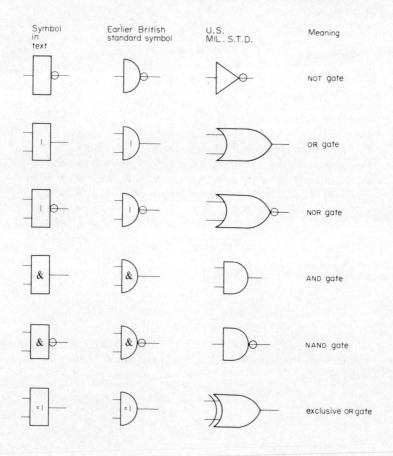

Figure B.1

APPENDIXES

These symbols are similar in many ways to the U.S. I.E.E.E. No. 91—ASA Y 32.14—1962 uniform shape symbols. Figure B.1 shows the symbols used in the text for the common logic gates together with two alternatives including the very widely used U.S. MIL. STD.

In the symbols shown in figure B.1, negation is indicated by a small circle at the appropriate point. The NAND and NOR gate symbols therefore differ from the symbols for AND and OR gates in that they have small circles shown at the gate outputs. Negation at the input of a gate can be indicated also, as shown in figure B.2 which represents a gate with the truth table shown in table B.1.

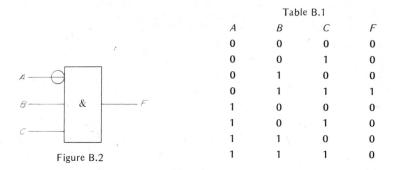

Table B.1

A	B	C	F
0	0	0	0
0	0	1	0
0	1	0	0
0	1	1	1
1	0	0	0
1	0	1	0
1	1	0	0
1	1	1	0

Figure B.2

In the case of bistable elements, the author has again chosen to adopt the shape used in the current British Standard. Figure B.3 shows the British Standard symbol for an R-S bistable which is almost identical to the symbol used in the text.
In the case of other bistable elements, symbols such as the one shown in figure B.4 have been used which retain the essential shape of the B.S. symbol with letters at the inputs and outputs to identify the particular operation. Direct-set and reset inputs are shown entering at the top and bottom of the bistable element symbol and the clock input if present is shown entering at the junction between the two parts of the symbol. These symbols are easily understood and have the advantage that they are sufficiently close to both European and U.S. standard symbols to enable the reader to follow directly diagrams which use the alternative symbols.

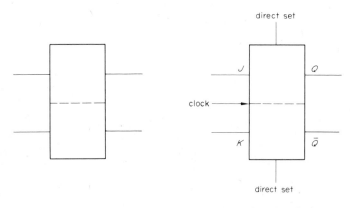

Figure B.3   Figure B.4

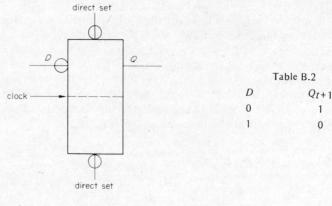

Figure B.5

Table B.2

D	Q_{t+1}
0	1
1	0

Negation at the input or output of a bistable element is indicated in the same way as for a combinational logic element symbol. Figure B.5 shows the symbol for a type D flip-flop in which the inverse of the D input is transferred to the Q output when a clock pulse is applied. The 'normal' state of the direct set and direct-reset inputs is 1. Taking direct-set or direct reset input to 0 will initiate the direct-set or reset operation.

Reference

B1 *B.S. News* (July 1971) 25.
B2 B.S. 3939 Section 21 Issue 1 (March 1969).

Index

Adder, carry bypass 101
    full 97
    half 96
    multidigit 99
    serial 102
Addition, binary 135
    octal 134
AND gate 11
AND–OR relationship 12
Arithmetic operation 96
Associativity rule 21
Asynchronous counter 63
Asynchronous system 2

BCD (binary coded decimal) numbers 138
Binary addition 135
Binary arithmetic 135
Binary division 136
Binary multiplication 136
Binary number system 135
Binary subtraction 136
Bistable, J–K 55
    R–S 50
    type D 54
Bistable circuit, discrete component 52
Bistable elements, integrated circuit 57
Bistable system 50
Boole, George 17
Boolean algebra 17
Boolean algebra expression 19
Boolean algebra relationship 21
Boolean algebra rules 21

Combination lock, electronic 83
Combinational logic circuit 2
Commutativity rule 21
Complement, generation of 104
    ones 136
    twos 136
Complementary MOS logic circuit 128
Counter 60
    four-stage 73
    ripple through 63
    scale of five 66
    scale of six 67
    scale of twelve 68
    synchronous 71
    two-stage 71
    variable sequence 78

Counting element 54

DTL (diode transistor logic) 125
De Morgan's theorem 21
Decimal number system 133
Decoder, integrated circuit 91
Decoding 91
Delay 121
Digital circuit parameters 120
Distributivity rule 21
Division, binary 136
    octal 109
Don't care situation 46
Dual in-line package 119

ECL (emitter coupled logic) 125
Equivalence system 30
Exclusive OR operation 16, 97

Fall time 121
Family, integrated circuit 118
Fan out 121
Flatpack 118
Flip-flop; see Bistable element
Floating point number 104
Frequency division 63

Gate, AND 11
    NAND 7
    NOR 7
    NOT 3
    OR 11
Gates, interconnection of 113

Hazard 71
High-speed TTL 124

Integrated circuits 118

Karnaugh map 35
    cell arrangement 36
    cell looping 36
    for don't care situation 47
    for NAND–NOR logic 41

Latch, bistable 54
Lift control system 85
Lock circuit 84
Logic convention 5

143

INDEX

Logic level 120
Logic probe 119
Low power TTL 124

Mantissa 104
Map methods 34
Master–slave technique 55
Minimisation 34
MOS logic circuit 126
Motor control system 24
    octal 134
Multiplication, binary 136
Multiplication system 107

NAND operation 7
NAND-NOR relationship 10
NOR operation 8
NOT operation 3
Negative logic 5
Noise, decoding 92
    immunity 6
    margin 6, 120
Numeration systems 132

OR operation 11
Octal numeration system 133

Package, integrated circuit 119
Patchboard 119
Positive logic 5
Power supply 120
Propagation delay 121

RTL (resistor–transistor logic) 122

Radix 133
    conversion of 136
Rate multiplier 113
Register 60
Restoring division 110
Ripple-carry adder 99
Rise time 121
Ripple through counter 63

Safety interlock system 24
Schottky clamped TTL 124
Serial addition system 102
Serial data transfer 62
Shift register 61
Signed binary numbers 103
Storage register 60
Subtraction 106, 137
Symbols for logic elements 140
Synchronous counter 70
Synchronous counter system 2

TTL (transistor–transistor logic) 124
Threshold level 120
Timing sequence 88
Tristate logic 125
Truth table 5

Up–down counter 78

Variable, boolean 17
Veitch map form 36
Venn map form 36

Wired logic 122